AF595333

Porous Materials for Environmental Applications

Porous Materials for Environmental Applications

Special Issue Editors

Antonio Gil

Miguel A. Vicente

MDPI • Basel • Beijing • Wuhan • Barcelona • Belgrade • Manchester • Tokyo • Cluj • Tianjin

Special Issue Editors
Antonio Gil
Departamento de Ciencias,
Universidad Pública de Navarra,
Edificio de los Acebos,
Campus de Arrosadía
Spain

Miguel A. Vicente
GIR-QUESCAT, Departamento
de Química Inorgánica,
Universidad de Salamanca
Spain

Editorial Office
MDPI
St. Alban-Anlage 66
4052 Basel, Switzerland

This is a reprint of articles from the Special Issue published online in the open access journal *Materials* (ISSN 1996-1944) (available at: https://www.mdpi.com/journal/materials/special_issues/porous_mat_environ_appl).

For citation purposes, cite each article independently as indicated on the article page online and as indicated below:

LastName, A.A.; LastName, B.B.; LastName, C.C. Article Title. *Journal Name* **Year**, *Article Number*, Page Range.

ISBN 978-3-03936-274-5 (Hbk)
ISBN 978-3-03936-275-2 (PDF)

Contents

About the Special Issue Editors

Antonio Gil (Full Professor of Chemical Engineering, Universidad Pública de Navarra, Spain): Professor Gil earned his BS and MS in Chemistry at University of Basque Country (San Sebastián) and his PhD in Chemical Engineering at University of Basque Country (San Sebastián). He did postdoctoral research at the Université catholique de Louvain (Belgium) working on Spillover and Mobility of Species on Catalyst Surfaces.

Miguel A. Vicente (Full Professor of Inorganic Chemistry, Universidad de Salamanca, Spain): Professor Vicente earned his BS and MS in Chemistry at Universidad de Salamanca, and his PhD in Inorganic Chemistry at UNED University (Madrid). He carried out postdoctoral research at the Université Pierre et Marie Curie (Paris, France) and at the Université catholique de Louvain (Belgium).

Communication

Spherical Activated Carbons with High Mechanical Strength Directly Prepared from Selected Spherical Seeds

Ana Amorós-Pérez, Laura Cano-Casanova, Mohammed Ouzzine, Mónica Rufete-Beneite, Aroldo José Romero-Anaya, María Ángeles Lillo-Ródenas * and Ángel Linares-Solano

MCMA Group, Department of Inorganic Chemistry and Materials Institute, University of Alicante, E-03080 Alicante, Spain; ana.amoros@ua.es (A.A.-P.); laura.cano@ua.es (L.C.-C.); ouzzine_mohamed@yahoo.fr (M.O.); monica.rufete@ua.es (M.R.-B.); ajromero@ua.es (A.J.R.-A.); linares@ua.es (A.L.-S.)

* Correspondence: mlillo@ua.es; Tel.: +34-965-90-35-45; Fax: +34-965-90-34-54

Received: 21 March 2018; Accepted: 4 May 2018; Published: 10 May 2018

Abstract: In the present manuscript, the preparation of spherical activated carbons (SACs) with suitable adsorption properties and high mechanical strength is reported, taking advantage of the retention of the spherical shape by the raw precursors. An easy procedure (carbonization followed by CO_2 activation) has been applied over a selection of three natural seeds, with a well-defined spherical shape and thermal stability: *Rhamnus alaternus* (RA), *Osyris lanceolate* (OL), and *Canna indica* (CI). After the carbonization-activation procedures, RA and CI, maintained their original spherical shapes and integrity, although a reduction in diameter around 48% and 25%, respectively, was observed. The porosity of the resulting SACs could be tuned as function of the activation temperature and time, leading to a spherical activated carbon with surface area up to 1600 m^2/g and mechanical strength similar to those of commercial activated carbons.

Keywords: spherical seeds; spherical activated carbons; activation; microporosity; mechanical properties

1. Introduction

Spherical activated carbons (SACs) are very interesting materials, which are attracting great attention because of their outstanding physical properties, such as wear resistance, mechanical strength, good adsorption performance, purity, low ash content, smooth surface, good fluidity, good packaging, low pressure drop, high bulk density, high micropore volume and tunable pore size distribution [1–5]. All these features make SACs suitable for various applications like blood purification, catalysts support, chemical protective clothing [2,6,7], in adsorption processes; both in gas phase (e.g., toluene, CO_2, CH_4 and H_2) [5,8–10] and solution (e.g., phenol) [11], as supercapacitors [12,13], in medicine for poison adsorption in living organisms [14], as catalyst supports for hydrogenation reactions [15,16], etc.

SACs can be prepared using several methods: by polymerization reactions [17], by agglomeration from mixtures of resin and activated carbon [18] or by hydrothermal synthesis [19–23]. All these methods imply the use of expensive or synthetic precursors (such as aerogels [17], divinylbenzene-derived polymers [24] and urea/formaldehyde resin [25]). However, nowadays it is common to look for cheaper precursors, such as coals [4], lignocellulosic materials [19–21] and carbohydrates [22,26–29].

Herein we present the preparation of SACs with high mechanical strength and tunable porosity from spherical seeds using an easy, cheap and a well-known method. This simple route allows the valorization of inexpensive and available biomass precursors, such as not edible seeds, to convert them in potentially useful and valuable materials like SACs. In particular, we focused our interest on the selected seeds that combine simultaneous spherical shape and thermal stability, and cover a wide range of diameters (from 1 to 7 mm). It should be highlighted that the size of the final materials would depend on the size of the used precursor. Among the tested spherical seeds, which accomplish these requirements (Table 1), the study was focused on three of them: *Rhamnus alaternus* (RA), *Osyris lanceolate* (OL), and *Canna indica* (CI) (Figure 1).

Table 1. Common and scientific names of the selected spherical seeds and their diameters.

Common Name	Scientific Name	Mean Diameter (mm)
Poppy	*Papaver rhoeas*	1
Amaranth	*Amaranthus hypochondriacus*	1
Millet	*Panicum miliaceum*	2
Mustard	*Sinapis alba*	3
Black pepper	*Piper nigrum*	4
False pepper	*Schinus molle*	4
Palm	*Phoenix dactylifera*	5
Indian shot	*Canna indica*	5
African sandalwood	*Osyris lanceolate*	5
Phoenicean juniper	*Juniperus phoenicea*	6
Mediterranean buckthorn	*Rhamnus alaternus*	7
Prickly juniper	*Juniperus oxycedrus*	7

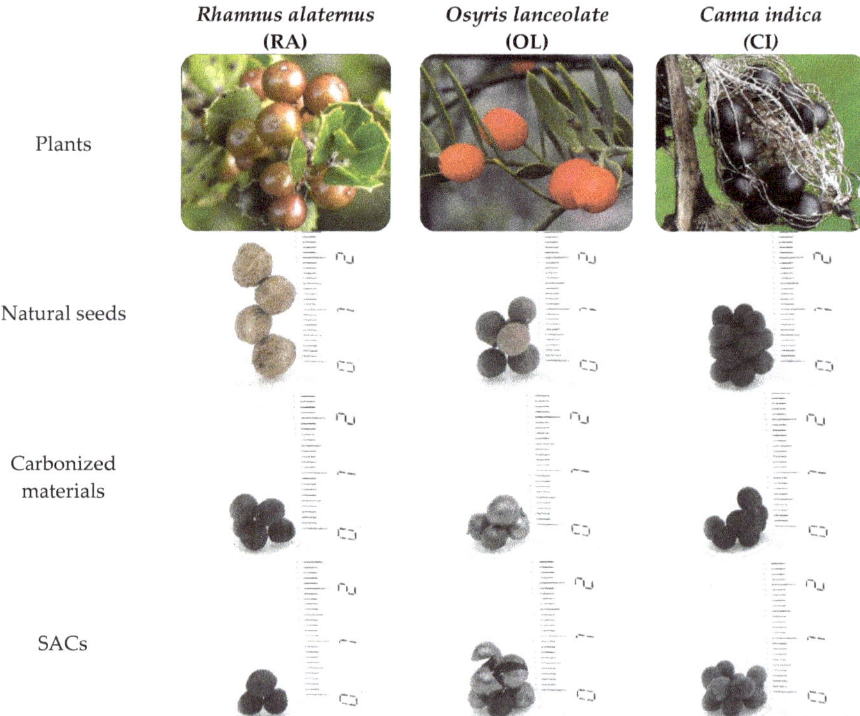

Figure 1. Natural seeds used as precursors for SACs preparation, together with the carbonized and activated spherical materials prepared from them.

2. Materials and Methods

2.1. Methodology

2.1.1. Carbonization Process

All the seeds were initially dried in an oven at 110 °C for 3 h. For the carbonization process, 2 g of such dried seeds were heated up to 850 °C, held for 2 h, in a horizontal furnace with N_2 flow of 300 mL/min, and a heating rate of 5 °C/min. The corresponding carbonization yields are summarized in Table 2.

Table 2. Mechanical strength (expressed as the percentage of the remaining mass after sieving, SRM%, see Section 2.2.3.) of the natural seeds and carbonization yields, textural properties and mechanical strength (SRM%) of the materials after carbonization.

Precursor	SRM [a] (%)	Yield [b] (%)	V_{DR} (N_2) [c] (cm^3/g)	V_{DR} (CO_2) [d] (cm^3/g)	SRM [e] (%)
RA	99.1	30	0.01	0.18	98.8
OL	99.9	21	0.01	0.19	99.4
CI	99.0	22	0.02	0.20	98.7

[a] SRM, remaining mass after sieving the natural spherical seeds, as percentage. [b] Yield, yield of carbonization process, as percentage. [c] V_{DR} (N_2), total micropore volume, obtained applying the Dubinin-Raduskevich method to data of N_2 adsorption isotherm at −196 °C. [d] V_{DR} (CO_2), narrow micropore volume, obtained applying the Dubinin-Raduskevich method to data of CO_2 adsorption isotherm at 0 °C. [e] SRM, remaining mass after sieving the carbonized materials, as percentage.

2.1.2. Activation Process

Carbonized seeds were activated using CO_2 in order to develop their porosity, using a CO_2 flow of 80 mL/min. To study the effect of temperature and time on the activation process, the samples were heated at 5 °C/min up to different temperatures: 800, 850, or 880 °C, and such temperatures were maintained for various fixed times, as described in Table 3.

Table 3. Activation conditions, activation percentages, SRM values and textural properties of some activated samples.

Precursor	T (°C)	t (h)	Burn-off (%)	S_{BET} [a] (m^2/g)	V_{DR} (N_2) [b] (cm^3/g)	V_{DR} (CO_2) [c] (cm^3/g)	V_{meso} [d] (cm^3/g)	SRM [e] (%)	$V_{N2}-V_{CO2}$ [g] (cm^3/g)
RA	800	10	6	492	0.20	0.25	0.03	NM[f]	< 0
	800	30	26	812	0.28	0.36	0.02	97.8	< 0
	800	40	33	889	0.40	0.33	0.03	NM[f]	0.07
	850	10	33	874	0.39	0.37	0.02	NM[f]	0.02
CI	800	5	33	856	0.39	0.35	0.05	94.9	0.04
	880	3	89	1616	0.64	0.37	0.19	85.3	0.27

[a] S_{BET}, BET surface area, obtained applying the BET method to data of N_2 adsorption isotherm at −196 °C. [b] V_{DR} (N_2), total micropore volume, obtained applying the Dubinin–Raduskevich method to data of N_2 adsorption isotherm at −196 °C. [c] V_{DR} (CO_2), narrow micropore volume, obtained applying the Dubinin–Raduskevich method to data of CO_2 adsorption isotherm at 0 °C. [d] V_{meso}, mesopore volume, obtained from N_2 adsorbed as liquid at $P/Po = 0.9$ minus the adsorbed volume at $P/P_0 = 0.2$ [30]. [e] SRM, remaining mass after sieving the activated materials, as percentage. [f] NM: not measured. [g] $V_{N2}-V_{CO2}$, difference between V_{DR} (N_2) and V_{DR} (CO_2).

2.2. Characterization

2.2.1. Morphology

Morphology of the original, carbonized and activated samples was characterized by Scanning Electron Microscopy (SEM) in a JSM-840 microscope (JEOL, Tokyo, Japan) with a scintillator–photomultiplier type secondary electron detector.

2.2.2. Surface Area and Pore Volumes

Textural characterization of precursors, carbonized and activated materials was performed using N_2 adsorption at −196 °C [31] and CO_2 at 0 °C [32] in a volumetric Autosorb-6B apparatus from Quantachrome. Before analysis, the samples were degassed at 250 °C for 4 h. The BET equation was applied to the nitrogen adsorption isotherm in the low-pressure region (relative pressure between 0.05–0.25) to get the apparent BET surface area, S_{BET} [30]. The Dubinin–Radushkevich equation was applied to the nitrogen adsorption isotherm to determine the total micropore volume (V_{DR} (N_2) corresponding to micropores of size below 2 nm) and to the carbon dioxide adsorption isotherms to determine narrow micropore volume (V_{DR} (CO_2), corresponding to micropores of size below 0.7 nm) [33]. Mesopore volume, V_{meso}, corresponding to pores between 2 and 20 nm, was estimated from N_2 adsorbed as liquid at P/P_0 = 0.9 minus the volume adsorbed at P/P_0 = 0.2 [30]. The difference between V_{DR} (N_2) and V_{DR} (CO_2) was calculated as an estimation of the micropore size distribution [30,31].

2.2.3. Mechanical Properties

The mechanical strength, defined as SRM%, was estimated by a method developed in our laboratory that consists of the evaluation of the sample mass remaining in a sieve after vigorous shaking (Figure 2). Thus, for each test, a known quantity of material was put in a cylindrical vial together with 15 stainless steel balls (Figure 2a). These vials were placed horizontally in a polymer mold used as immobilizing support (Figure 2b), which was placed in an electromagnetic sieve shaker CIPSA RP08 Ø200/203 for 20 min at power number 8 (equivalent to the shaking speed of 1.8 mm of vibration amplitude per second) (Figure 2c). Then, the samples were sieved using a sieve (300 μm) and the resulting residue (the sample not converted to dust in the sieving step) was weighed (Figure 2d). The mechanical strength was expressed as the percentage of the remaining mass after sieving (SRM%). The validation of this method was performed by analyzing the mechanical properties of several commercial activated carbons (Table 4), and such values were used as reference to confirm that the mechanical properties of our SACs are similar to those of their commercial counterparts. Note that the SRM values for the selected commercial ACs are in the range between 72% and 97%. The analysis of the mechanical properties was performed for the precursors, carbonized materials (Table 2) and for the activated ones (Table 3).

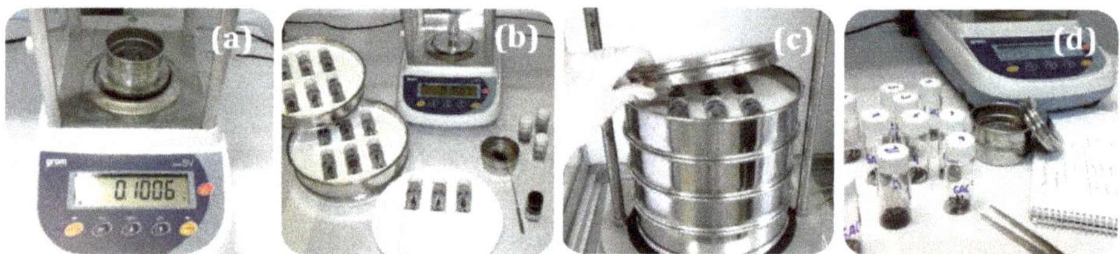

Figure 2. Depiction of materials and procedure for the mechanical strength evaluation of the samples: (**a**) a weighted sample was sieved using in 300 μm sieve and put in a vial; (**b**) 15 steel balls were also incorporated in the vial, which was then placed in a polymer mold; (**c**) the molds were placed in the sieve shaker during 20 min; (**d**) the sample was sieved again in a 300 μm sieve and the residue (the sample not converted to dust in the sieving step) was collected and weighed to calculate the sample remaining mass percentage (SRM%).

Table 4. Textural properties and SRM values of some commercial activated carbons.

Name	Commercial Name	Morphology and Size	S_{BET} [a] (m^2/g)	V_{DR} (N_2) [b] (cm^3/g)	V_{DR} (CO_2) [c] (cm^3/g)	V_{meso} [d] (cm^3/g)	SRM (%)
CW	Mead Westvaco, WVA1100	Granular (10 × 25 mesh)	1796	0.72	0.34	0.42	72
CK	Kureha Corporation carbon from petroleum pith	Spherical (0.75 μm)	1185	0.57	0.42	0.02	97
ROX	NORIT® ROX	Pellets (0.8 mm)	1354	0.60	0.40	0.07	92

[a] $S_{BET,}$ BET surface area, obtained applying the BET method to data of N_2 adsorption isotherm at −196 °C. [b] V_{DR} (N_2), total micropore volume, obtained applying the Dubinin-Raduskevich method to data of N_2 adsorption isotherm at −196 °C. [c] V_{DR} (CO_2), narrow micropore volume, obtained applying the Dubinin–Raduskevich method to data of CO_2 adsorption isotherm at 0 °C. [d] V_{meso}, mesopore volume, obtained from N_2 adsorbed as liquid at P/P_0 = 0.9 minus the adsorbed volume at P/Po = 0.2 [30].

3. Results

Figure 3 shows that the retention of the desired original spherical shape during the carbonization process was achieved for these three seeds, though this step generally led to a decrease in the diameter of the materials. Such variation depends on the type of seed: for RA the size was significantly reduced (around 3 mm, which represents 40% reduction), while CI and OL were only slightly shortened (in both cases about 1 mm, around 20%). This could be related with the intrinsic natural differences in the composition of the seeds.

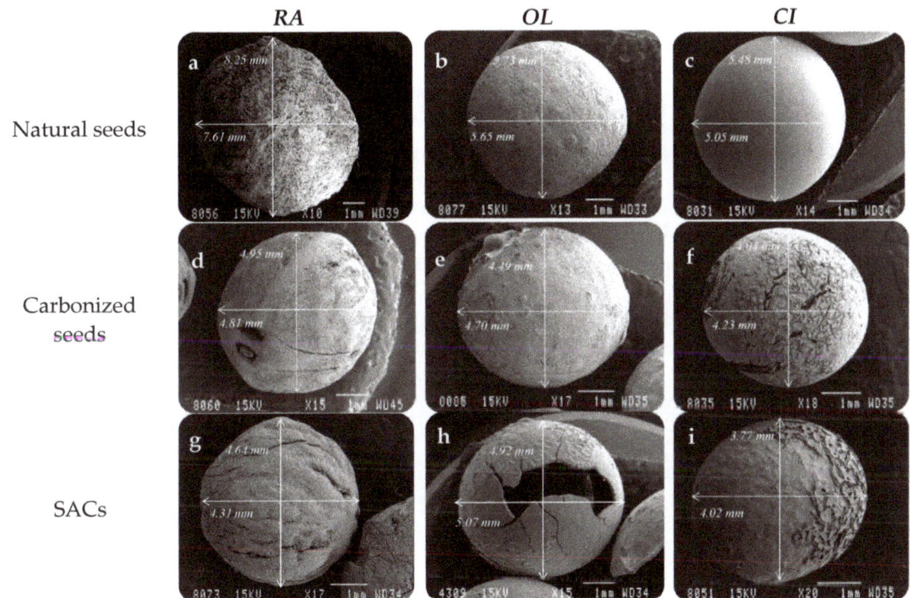

Figure 3. SEM images of precursors, carbonized materials and final SACs. Experimental preparation conditions of the materials: (**a–c**) dried at 110 °C for 3 h; (**d–f**) carbonized at 850 °C for 2 h in 300 mL/min N_2 flow; (**g**) activated at 800 °C for 30 h using 80 mL/min CO_2 flow; (**h**) activated at 800 °C for 2 h in 80 mL/min CO_2 flow; (**i**) activated at 800 °C for 5 h in 80 mL/min CO_2 flow.

Table 2 reports carbonization yields and values of mechanical properties (SRM%) for CI, RA, and OL carbonized seeds, together with SRM values for the precursors, as reference. It shows that: (i) the carbonization yields ranged from 21 to 30%, which is in the range of typical values expected for lignocellulosic materials [34]; (ii) micropore volumes determined by CO_2 adsorption were larger than those measured by N_2, indicating that the mean micropore sizes were below 0.7 nm [31], and (iii) SRM values were higher than 98.7%, indicating that the samples possess high mechanical resistance. Such porous texture, together with the mechanical properties, made these carbonized materials potentially useful as spherical carbon molecular sieves.

The conditions of the activation process were also optimized in order to obtain similar burn-off percentage (around 30%) and maintain the spherical morphology. Figure 3 shows that RA and CI seeds retained their original shape after activation, and their sizes were minimally affected (around 0.5 mm (8% reduction with respect to the size) before the activation, and 0.2 mm (5%), respectively). Only OL seeds were broken after activation, and this occurred for all the explored activation conditions. Hence, from OL only spherical carbonized materials could be prepared. It is important to mention that the carbonized and activated materials from both RA and CI remained physically intact (without cracks) and the same occurred for carbonized OL, whereas only the material obtained from OL, after the activation process showed cracks, that can be visually distinguished.

With respect to the activation yields (Table 3), CI was the more reactive candidate, since a shorter activation time was required to get 33% burn-off. For the RA precursor, as expected, the burn-off percentage at constant temperature increases proportionally with the reaction time. Interestingly, the desired activation percentage (33%) could also be achieved for RA using higher temperatures (850 °C instead of 800 °C) and shorter times (10 h instead of 40 h).

Regarding the textural properties, Table 3 shows that, for RA seeds, although there exists a linear relationship between the activation time and the burn-off percentage, no direct correlation was found when analyzing the effect of the activation time on the porosity development. For this precursor, the same burn-off percentage, 33%, and similar surface area values (about 880 m^2/g) have been obtained using different combinations of activation temperature and time. Low activation times (10 and 30 h) led to materials with the mean micropore sizes below 0.7 nm, whereas the micropore size was around that value for larger activation time and temperature.

Similar experimental conditions screening for CI precursor highlighted that higher adsorption capacities can be developed from it, which reached 1616 m^2/g when treating up to 880 °C for 3 h. For the CI activated material with surface area above 1600 m^2/g, the fact that total micropore volume determined by N_2 adsorption is much larger than that measured by CO_2, is indicative of the average pore size above 0.7–1 nm [30].

By comparing SRM values for natural and carbonized materials in Table 2, it can be observed that natural precursors show slightly higher mechanical strength than the corresponding carbonized spheres. Table 3 contains the SRM values for SACs, indicating that their mechanical properties are only slightly reduced after the activation process. This is probably due to the high number of heteroatoms linked to the carbon material, and eliminated during the activation [35]. However, SRM values are in the range of 95% for samples with areas around 800 m^2/g, and about 85% when the BET surface area surpasses 1600 m^2/g, which indicated that the samples generally display significant mechanical properties that lie in the range of the selected common commercial references (CW, CK and ROX) (Table 4 and Figure 4).

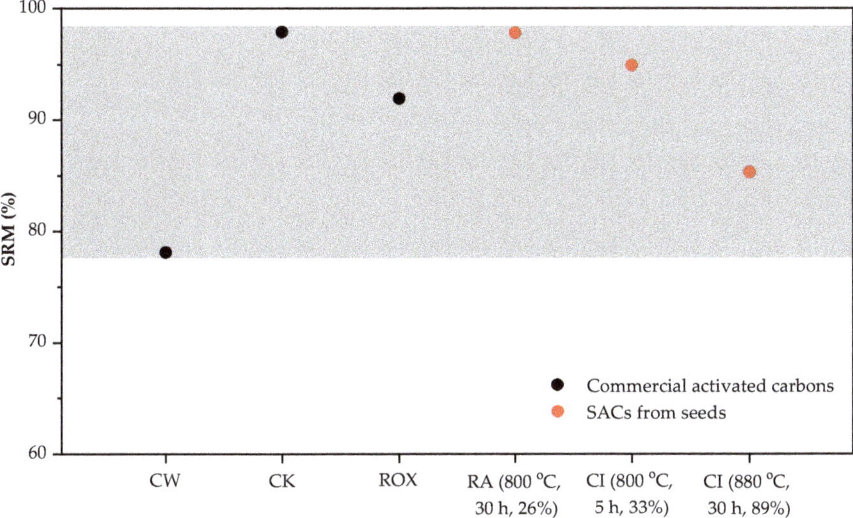

Figure 4. SRM values for commercial activated carbons (black color) and for some activated seeds prepared in this work (orange color).

4. Conclusions

In this work spherical activated carbons with high mechanical strength and well-developed porosity were prepared while maintaining the spherical shape of the natural seeds, selected as carbon precursors. The three reported candidates: RA, OL and CI, could be successfully converted into spherical activated carbon materials using a well-known, simple and cheap method and, additionally, CI and RA maintained their original spherical shapes and integrity all along the activation process, avoiding breakage. Their diameter sizes were notably reduced after the carbonization step and only slightly affected by the activation process. The mechanical properties for all the activated materials were found to be similar to those of common commercial activated carbons with different morphologies (granular, spherical and pellets). Interestingly, depending on the precursor and/or on the activation conditions, significant differences in porosity development and micropore size distributions were obtained, reaching specific surface areas up to 1600 m^2/g. The interesting properties of the prepared materials, together with their spherical morphology, make them interesting candidates for many applications.

Author Contributions: Á.L.-S. conceived and designed the experiments; A.A.-P., L.C.-C., M.O., M.R.-B. and A.J.R.-A. performed the experiments; A.A.-P., L.C.-C., M.O., M.A.L.-R. and Á.L.-S. analyzed the data; A.A.-P., M.O., A.J.R.-A., M.A.L.-R. and Á.L.-S. contributed reagents/materials/analysis tools; A.A.-P., A.J.R.-A., M.A.L.-R. and Á.L.-S. wrote the paper.

Funding: This research received no external funding.

Acknowledgments: The authors thank the Spanish Ministry of Economy and Competitiveness (MINECO) and FEDER, project of reference CTQ2015-66080-R, GV/FEDER (PROMETEOII/2014/010) and University of Alicante (VIGROB-136) for financial support. Pilar Garcia Cardona, from Universidad Nacional de Colombia-Medellin is acknowledged for providing some of the spherical materials.

Conflicts of Interest: The authors declare no conflict of interest.

References

1. Liu, J.; Wickramaratne, N.P.; Qiao, S.Z.; Jaroniec, M. Molecular-based design and emerging applications of nanoporous carbon spheres. *Nat. Mater.* **2015**, *14*, 763–774. [CrossRef] [PubMed]
2. Liu, Z.; Ling, L.; Qiao, W.; Liu, L. Effect of hydrogen on the mesopore development of pitch-based spherical activated carbon containing iron during activation by steam. *Carbon* **1999**, *37*, 2063–2066. [CrossRef]
3. Liang, X.; Zhang, R.; Liu, C.; Liu, X.; Qiao, W.; Zhan, L.; Ling, L. Preparation of polystyrene-based activated carbon spheres and their adsorption of dibenzothiophene. *New Carbon Mater.* **2009**, *9*, 8–13. [CrossRef]
4. Gryglewicz, G.; Grabas, K.; Lorenc-Grabowska, E. Preparation and characterization of spherical activated carbons from oil agglomerated bituminous coals for removing organic impurities from water. *Carbon* **2002**, *40*, 2403–2411. [CrossRef]
5. Romero-Anaya, A.J.; Lillo-Ródenas, M.A.; Linares-Solano, A. Spherical activated carbons for low concentration toluene adsorption. *Carbon* **2010**, *48*, 2625–2633. [CrossRef]
6. Yenisoy-Karakaş, S.; Aygün, A.; Güneş, M.; Tahtasakal, E. Physical and chemical characteristics of polymer-based spherical activated carbon and its ability to adsorb organics. *Carbon* **2004**, *42*, 477–484. [CrossRef]
7. Lee, C.; Hsu, S. Preparation of spherical encapsulation of activated carbons and their adsorption capacity of typical uremic toxins. *J. Biomed. Mater. Res.* **1990**, *24*, 243–258. [CrossRef] [PubMed]
8. Wickramaratne, N.P.; Jaroniec, M. Activated carbon spheres for CO_2 adsorption. *Appl. Mater. Interfaces* **2013**, *5*, 1849–1855. [CrossRef] [PubMed]
9. Romero-Anaya, A.J.; Lillo-Ródenas, M.A.; Linares-Solano, A. Activation of a spherical carbon for toluene adsorption at low concentration. *Carbon* **2014**, *77*, 616–626. [CrossRef]
10. Li, Y.; Li, D.; Rao, Y.; Zhao, X.; Wu, M. Superior CO_2, CH_4, and H_2 uptakes over ultrahigh-surface-area carbon spheres prepared from sustainable biomass-derived char by CO_2 activation. *Carbon* **2016**, *105*, 454–462. [CrossRef]
11. Liu, C.; Liang, X.; Liu, X.; Wang, Q.; Teng, N.; Zhan, L.; Zhang, R.; Qiao, W.; Ling, L. Wettability modification of pitch-based spherical activated carbon by air oxidation and its effects on phenol adsorption. *Appl. Surf. Sci.* **2008**, *254*, 2659–2665. [CrossRef]
12. Xuan, H.; Wang, Y.; Lin, G.; Wang, F.; Zhou, L.; Dong, X.; Chen, Z. Air-assisted activation strategy for porous carbon spheres to give enhanced electrochemical performance. *RSC Adv.* **2016**, *6*, 15313–15319. [CrossRef]
13. Sevilla, M.; Fuertes, A.B. A green approach to high-performance supercapacitor electrodes: The chemical activation of hydrochar with potassium bicarbonate. *ChemSusChem* **2016**, 1880–1888. [CrossRef] [PubMed]
14. Tsivadze, A.Y.; Gur'yanov, V.V.; Petukhova, G.A. Preparation of spherical activated carbon from furfural, its properties and prospective applications in medicine and the national economy. *Prot. Met. Phys. Chem. Surf.* **2011**, *47*, 612–620. [CrossRef]
15. Bo, B.; Wasserscheid, P.; Etzold, B.J. Polymer-based spherical activated carbon as easy-to-handle catalyst support for hydrogenation reactions. *Chem. Eng. Technol.* **2016**, *39*, 276–284. [CrossRef]
16. Rufete-Beneite, M.; Román-Martínez, M.C.; Linares-Solano, A. Insight into the immobilization of ionic liquids on porous carbons. *Carbon* **2014**, *77*, 947–957. [CrossRef]
17. Long, D.; Zhang, R.; Qiao, W.; Zhang, L.; Liang, X.; Ling, L. Biomolecular adsorption behavior on spherical carbon aerogels with various mesopore sizes. *J. Colloid Interface Sci.* **2009**, *331*, 40–46. [CrossRef] [PubMed]
18. Oh, W.C.; Kim, J.G.; Kim, H.; Chen, M.L.; Zhang, F.J.; Zhang, K.; Meng, Z. Da Preparation of spherical activated carbon and their physicochemical properties. *J. Korean Ceram. Soc.* **2009**, *46*, 568–573. [CrossRef]
19. Luo, G.; Shi, W.; Chen, X.; Ni, W.; Strong, P.J.; Jia, Y.; Wang, H. Hydrothermal conversion of water lettuce biomass at 473 or 523 K. *Biomass Bioenergy* **2011**, *35*, 4855–4861. [CrossRef]
20. Hoekman, S.K.; Broch, A.; Robbins, C. Hydrothermal Carbonization of Lignocellulosic Biomass. *Energy Fuels* **2014**, *25*, 1802–1810. [CrossRef]
21. Sevilla, M.; Maciá-Agulló, J.A.; Fuertes, A.B. Hydrothermal carbonization of biomass as a route for the sequestration of CO_2: Chemical and structural properties of the carbonized products. *Biomass Bioenergy* **2011**, *35*, 3152–3159. [CrossRef]
22. Sun, X.; Li, Y. Ga_2O_3 and GaN semiconductor hollow spheres. *Angew. Chem. Int. Ed.* **2004**, *43*, 3827–3831. [CrossRef] [PubMed]

23. Bedin, K.C.; Cazetta, A.L.; Souza, I.P.A.F.; Pezoti, O.; Souza, L.S.; Souza, P.S.C.; Yokoyama, J.T.C.; Almeida, V.C. Porosity enhancement of spherical activated carbon: Influence and optimization of hydrothermal synthesis conditions using response surface methodology. *J. Environ. Chem. Eng.* **2018**, *6*, 991–999. [CrossRef]
24. Zhu, Z.; Li, A.; Yan, L.; Liu, F.; Zhang, Q. Preparation and characterization of highly mesoporous spherical activated carbons from divinylbenzene-derived polymer by $ZnCl_2$ activation. *J. Colloid Interface Sci.* **2007**, *316*, 628–634. [CrossRef] [PubMed]
25. Wang, D.; Chen, M.; Wang, C.; Bai, J.; Zheng, J. Synthesis of carbon microspheres from urea formaldehyde resin. *Mater. Lett.* **2011**, *65*, 1069–1072. [CrossRef]
26. Yao, C.; Shin, Y.; Wang, L.Q.; Windisch, C.F.; Samuels, W.D.; Arey, B.W.; Wang, C.; Risen, W.M.; Exarhos, G.J. Hydrothermal dehydration of aqueous fructose solutions in a closed system. *J. Phys. Chem. C* **2007**, *111*, 15141–15145. [CrossRef]
27. Sevilla, M.; Fuertes, A.B. The production of carbon materials by hydrothermal carbonization of cellulose. *Carbon* **2009**, *47*, 2281–2289. [CrossRef]
28. Wang, F.L.; Pang, L.L.; Jiang, Y.Y.; Chen, B.; Lin, D.; Lun, N.; Zhu, H.-L.; Liu, R.; Meng, X.L.; Wang, Y.; et al. Simple synthesis of hollow carbon spheres from glucose. *Mater. Lett.* **2009**, *63*, 2564–2566. [CrossRef]
29. Romero-Anaya, A.J.; Ouzzine, M.; Lillo-Ródenas, M.A.; Linares-Solano, A. Spherical carbons: Synthesis, characterization and activation processes. *Carbon* **2014**, *68*, 296–307. [CrossRef]
30. Rodriguez-Reinoso, F.; Linares-Solano, A. *Microporous Structure of Activated Carbons as Revealed by Adsorption Methods*; Thrower, P.A., Ed.; Marcel Dekker Inc.: New York, NY, USA, 1989; Volume 21, ISBN 0-8247-7939-8.
31. Linares-Solano, Á.; Salinas-Martínez de Lecea, C.; Alcañiz-Monge, J.; Cazorla-Amorós, D. Further advances in the characterization of microporous carbons by physical adsorption of gases. *Tanso* **1998**, *185*, 316–325. [CrossRef]
32. Cazorla-Amorós, D.; Alcañiz-Monge, J.; Linares-Solano, A. Characterization of activated carbon fibers by CO_2 adsorption. *Langmuir* **1996**, *12*, 2820–2824. [CrossRef]
33. Lillo-Ródenas, M.A.; Marco-Lozar, J.P.; Cazorla-Amorós, D.; Linares-Solano, A. Activated carbons prepared by pyrolysis of mixtures of carbon precursor/alkaline hydroxide. *J. Anal. Appl. Pyrolysis* **2007**, *80*, 166–174. [CrossRef]
34. Rodríguez-Reinoso, F.; Molina-Sabio, M. Activated carbons from lignocellulosic materials by chemical and/or physical activation: An overview. *Carbon* **1992**, *30*, 1111–1118. [CrossRef]
35. Contreras, M.S.; Páez, C.A.; Zubizarreta, L.; Léonard, A.; Blacher, S.; Olivera-Fuentes, C.G.; Arenillas, A.; Pirard, J.P.; Job, N. A comparison of physical activation of carbon xerogels with carbon dioxide with chemical activation using hydroxides. *Carbon* **2010**, *48*, 3157–3168. [CrossRef]

Article

One-Step Hydrothermal Synthesis of Zeolite X Powder from Natural Low-Grade Diatomite

Guangyuan Yao, Jingjing Lei, Xiaoyu Zhang, Zhiming Sun * and Shuilin Zheng *

School of Chemical and Environmental Engineering, China University of Mining and Technology (Beijing), Beijing 100083, China; slxw.yao1990@hotmail.com (G.Y.); 18811568019@163.com (J.L.); zhangxiaoyukdbj@163.com (X.Z.)

* Correspondence: zhimingsun@cumtb.edu.cn (Z.S.); shuilinzheng8@gmail.com (S.Z.)

Received: 15 May 2018; Accepted: 25 May 2018; Published: 28 May 2018

Abstract: Zeolite X powder was synthesized using natural low-grade diatomite as the main source of Si but only as a partial source of Al via a simple and green hydrothermal method. The microstructure and surface properties of the obtained samples were characterized by powder X-ray diffraction (XRD), scanning electron microscopy (SEM), wavelength dispersive X-ray fluorescence (XRF), calcium ion exchange capacity (CEC), thermogravimetric-differential thermal (TG-DTA) analysis, and N_2 adsorption-desorption technique. The influence of various synthesis factors, including aging time and temperature, crystallization time and temperature, Na_2O/SiO_2 and H_2O/Na_2O ratio on the CEC of zeolite, were systematically investigated. The as-synthesized zeolite X with binary meso-microporous structure possessed remarkable thermal stability, high calcium ion exchange capacity of 248 mg/g and large surface area of 453 m^2/g. In addition, the calcium ion exchange capacity of zeolite X was found to be mainly determined by the crystallization degree. In conclusion, the synthesized zeolite X using diatomite as a cost-effective raw material in this study has great potential for industrial application such as catalyst support and adsorbent.

Keywords: diatomite; zeolite X; hydrothermal method; calcium ion exchange capacity

---

1. Introduction

Zeolites are crystalline aluminosilicates built from TO_4 tetrahedra (T = Si and Al) with excellent properties of high surface area, uniform and precise microporosity, shape selectivity, high ion-exchange capacity, strong Brønsted acidity and high thermal and hydrothermal stability [1]. Therefore, zeolites have been widely used in many environmental and other industrial applications, such as ion exchange [2–5], catalysts [6–9], membrane separations [10–12] and adsorbents [13–17].

The principle raw materials used for the synthesis of the zeolites are different sources of silica and alumina, which are usually composed of sodium silicates, sodium aluminate, aluminum salts or colloidal silica. However, traditional methods for synthesizing zeolites typically involve chemical reagents as starting materials or crystallization from a gel or clear solution under hydrothermal conditions, which have the disadvantages of high cost, excessive waste, and unfriendly nature to the environment. Therefore, many attempts are underway for economical synthesis of zeolites. In general, natural aluminosilicate and silicate minerals and industrial solid wastes have been explored as silica and/or alumina source because they are cost-effective precursors and can lead to reduction of the synthesis costs. Until now, There have been many studies on synthesizing zeolites from natural minerals such as kaolinite [18–20], bentonite [19], feldspar [19,21] and other precursors [22–26].

Although zeolites have been synthesized from the solid wastes, such as fly ash [27–29], rice husk ash [30] and coal gangue [22], the uncertainty in their supplies and the impurity in their components may limit their practical application. Therefore, direct synthesis of zeolites from natural aluminosilicate and silicate minerals with abundant reserves in the earth has been pursued because of its great potential in reducing the generation of hazardous wastes, saving energy, and possibly altering the properties of the resulting zeolites [31]. However, most aluminosilicate minerals are inactive, which restricted their practical application in the zeolite synthesis. Additionally, even after the thermal activation, only part of the aluminum-oxygen bonds can be broken, which means that just part of the Al_2O_3 and a small amount of SiO_2 only can participate in the zeolite synthesis.

Diatomite is an interesting silica material because of its relatively low cost, large reserves and its highly reactive amorphous state derived from silica skeletons of diatoms. Being silica rich, diatomite serves as the silica source in the synthesis of zeolites, which is cost-effective. In addition, diatomite is amorphous and highly reactive and therefore, unnecessary to transform it into a reactive state as done with many crystalline minerals [18–21]. Furthermore, parts of secondary building units of minerals could be preserved in the synthesis process. Therefore, it is necessary to explore detailed research on the preparation and formation mechanisms of zeolites from diatomite.

In the present work, we report on the synthesis of zeolite X from natural low-grade diatomite with high crystallization degree under hydrothermal conditions. Additionally, the effect of factors such as aging time and temperature, crystallization time and temperature, Na_2O/SiO_2 and H_2O/Na_2O ratio on the synthesis of zeolite in the SiO_2-Al_2O_3-Na_2O-H_2O system were systematically investigated. The crystallization degree of the products was evaluated by XRD, SEM and CEC analysis. Meanwhile, the formation mechanism was also explored and discussed.

2. Experimental

2.1. Materials

Diatomite (Dt) is obtained from Linjiang City, Jilin Province, China. Its main chemical composition by wt % is: SiO_2: 63.77%; Al_2O_3: 18.97%; Fe_2O_3: 1.48%; CaO: 0.48%; K_2O: 0.16%; Na_2O: 0.04%. It was grounded to a size smaller than 30 mesh and dried at 105 °C. Commercial zeolites X were purchased from Tianjin yuanli Reagent Co. (Tianjin yuanli, China). Sodium hydroxide, aluminum hydroxide (nordstrandite) and the other chemicals used in the experiments were purchased from Xilong Reagent Co. (Xilong, China). All chemicals were of analytical reagent grade and used without any further purification. Deionized water was used throughout this study.

2.2. Preparation of Zeolite X

The synthesis of zeolites from diatomite includes three processes as follows: gel formation, aging and crystallization. Initially, diatomite and $Al(OH)_3$ were dispersed in NaOH solution under vigorous magnetic stirring to form a homogeneous dispersion. The amount of diatomite and $Al(OH)_3$ were according to the 1.13 (molar ratio) of [Si/Al], and the amount of NaOH solution was according to the molar ratios of [Na_2O/SiO_2] and [H_2O/Na_2O]. Subsequently, the slurry was subjected to aging for 0–120 min at 30–60 °C. Then, the mixed solution was put into a Teflon-lined stainless steel autoclave. Finally, the container was closed and crystallized at 90–120 °C for 3–9 h. After that, the autoclave was cooled to room temperature naturally, and the samples were removed from the reactor, filtered, and washed with deionized water until the pH of the filtrate reached 6–7. Finally, the wet washed solids were dried overnight at 105 °C before further measurement and characterization.

2.3. Characterization

X-ray diffraction (XRD) analysis were performed on a D8 advance X-ray diffractometer (Bruker, Germany) equipped with Cu-Kα radiation (λ = 0.154056 nm) to identify the crystalline phase of the obtained X zeolite products. The samples were scanned in the 2θ range of 5° to 50° with a 0.02° step at a scanning speed of 4°/min. The surface morphology of the samples was observed by S-4800 scanning electron microscope (Hitachi, Japan). Nitrogen adsorption-desorption isotherms were measured at 77 K using an ASAP 2020 instrument (Micromeritics, Norcross, GA, USA), after evacuation of the samples at 350 °C for

4 h. The specific surface area (SBET) and microporous volume (Vmicro) were calculated using the BET and t-plot methods, respectively. Pore size distribution curves were calculated by Barrett-Joyner-Halenda (BJH) method. The crystallization behavior of zeolites as well as the thermal properties of the composites was monitored and evaluated using a Mettler TGA/DSC 1 SF/1382 equipment (NETZSCH, Germany). The TGA/DSC measurements were carried out in air flow with a heating rate of 5 °C/min from 25–900 °C. Chemical composition of the sample was determined using wavelength dispersive X-ray fluorescence spectrometry (XRF, Shimadzu, Japan) on a Shimadzu XRF-1800 apparatus. Calcium ion exchange capacity (CEC) was determined as follows [22]: Typically, 0.5 g of zeolite sample was poured into 500 mL of 0.005 M $CaCl_2$ solution and the mixture was shaken for 20 min at 35 °C. Then the filtrate was analyzed by the addition of calconcarboxylic acid and EDTA to determine the CEC of the samples.

3. Results and Discussion

3.1. Starting Materials

The XRD pattern (a) and SEM images (b,c) of raw diatomite are shown in Figure 1. According to the XRD pattern of diatomite, the broad reflection centered at 2θ = 15–30° was attributed to the amorphous silica, and the peaks at 2θ = 20.08° and 26.65° were ascribed to quartz. In addition, the peaks at 2θ = 11.88°, 27.35° and 35°–40° were characteristic to kaolinite-montmorillonite [32], which were the main Al source. As shown in Figure 1b,c, the diatomite exhibits highly porous cylinder-like or boat-like shape.

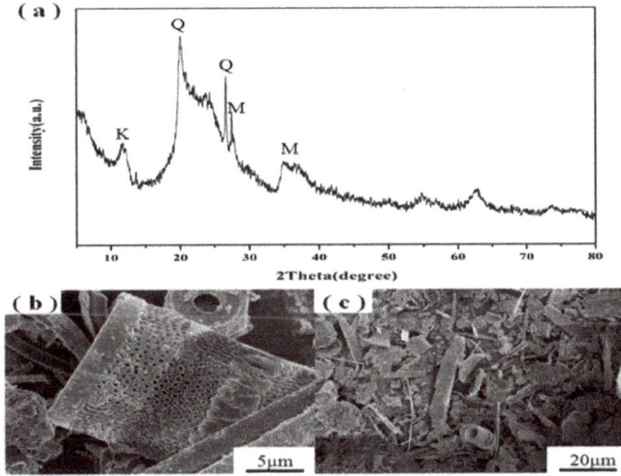

Figure 1. XRD pattern (**a**) and SEM images (**b**,**c**) of raw diatomite. Q = quartz; M = montmorillonite; K = kaolinite.

3.2. Effect of Crystallization

Crystallization conditions are important parameters that control the crystallization of zeolites. The crystallization conditions could also change the autogenous pressure in the autoclave and may alter the structure of the resulting zeolites. Then, a batch of experiments were carried out under different crystallization temperatures and times, while the other preparation conditions including aging temperature and time, Na_2O/SiO_2 and H_2O/Na_2O ratio were kept constant. Namely: 30 °C of aging temperature, 60 min of aging time, 40 of H_2O/Na_2O ratio and 3.0 of Na_2O/SiO_2 ratio. The XRD patterns of samples with different crystallization temperatures and times are shown in Figure 2a,b, respectively. It can be seen from Figure 2a that the sample at a crystallization temperature of 90 °C showed only one strong diffraction peak at 2θ = 18.35°, which can be attributed to aluminum hydroxide indicating that diatomite was dissolved in the alkaline solution, but the silica did not react with aluminum hydroxide. When the temperature was raised to 100 °C, typical diffraction peaks of X zeolite (JCPDS 38-0237) can be seen at 2θ = 6.10°, 9.97°, 11.76°, 15.39°, 18.46°, 20.12°, 23.24°, 26.58° and 30.86° suggesting that zeolite X started to crystallize (Figure 2a). However, the diffraction peak of aluminum hydroxide can also be seen at 2θ = 18.35°, which suggested an incomplete reaction of aluminum hydroxide and silica from diatomite. Highly crystalline single phase zeolite X was formed as the crystallization temperature was raised to 110 °C. When the reaction temperature was further raised to 120 °C, almost pure phase of zeolite X was observed along with some zeolite A (JCPDS 43-0142). It indicated that the higher crystallization temperature of 120 °C caused the transformation of some zeolite X into zeolite A. As shown in Figure 2b, no zeolite X was obtained at crystallization time of 3 h at 110 °C. With the increase of crystallization time, the crystallization degree of zeolite X was enhanced. However, longer reaction time caused the transformation of zeolite X into zeolite A at 110 °C. Therefore, suitable crystallization temperature and time are essential for the formation of high purity zeolite X.

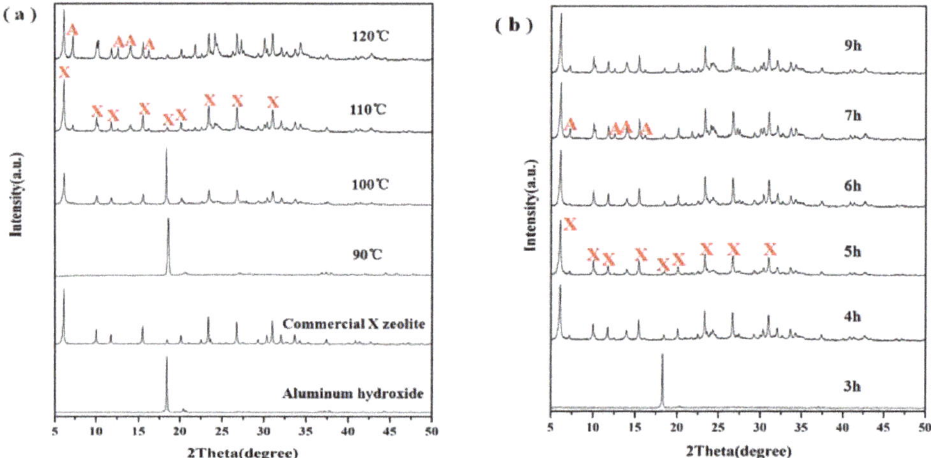

Figure 2. XRD patterns showing the effect of (**a**) crystallization temperature at crystallization time of 5 h and (**b**) crystallization time at crystallization temperature of 110 °C on zeolite formation.

SEM images of samples obtained at different crystallization temperatures are shown in Figure 3. When the crystallization temperature was 90 °C, the reactants formed spherical aggregates of ill-defined particles. With the increase of crystallization temperature, the products gradually formed regular crystals with smooth faces. However, the crystallization temperature at 110 °C formed better octahedral crystal morphology along with uniform distribution of crystals (Figure 3e,f). In addition, amorphous material was hardly noticeable in the SEM images of as-synthesized zeolite X suggesting the formation of highly crystalline zeolite X. When the crystallization temperature was increased up to 120 °C, some cubic crystals appeared (Figure 3g,h), which could be attributed to the formation of zeolite A.

The morphologies of the samples prepared at different crystallization times at 110 °C are shown in Figure 4. When the crystallization time was 4 h, the synthesized products formed crystals with no distinct faces, and the particles were not uniform. When the crystallization time reached 5 h, the synthesized products formed regular crystals with smooth faces. When the crystallization time was up to 6 h, the synthesized products do not show smooth faces. The morphologies as determined by SEM with different crystallization temperatures and times are in accordance with the XRD results. When the crystallization temperature and time were 110 °C and 5 h, the prepared samples showed higher crystallization degree.

The influence of crystallization temperature and time on the CEC is presented in Figure 5a,b. Increasing temperature of crystallization up to 110 °C increased the CEC but by further increasing the temperature up to 130 °C, the CEC of prepared samples gradually decreased (Figure 5a). Furthermore, the CEC of samples firstly increased with the increase of crystallization time and then gradually decreased when the time was over 5 h (Figure 5b). Combined with the analysis of SEM and XRD, it is concluded that the CEC increased because of the enhancement of crystallization degree of zeolite X.

Figure 3. SEM images of zeolites obtained with crystallization temperature at (**a**,**b**) 90 °C, (**c**,**d**) 100 °C, (**e**,**f**) 110 °C and (**g**,**h**) 120 °C.

Figure 4. SEM images of zeolites obtained with crystallization time at (**a**,**b**) 4 h, (**c**,**d**) 5 h and (**e**,**f**) 6 h at 110 °C.

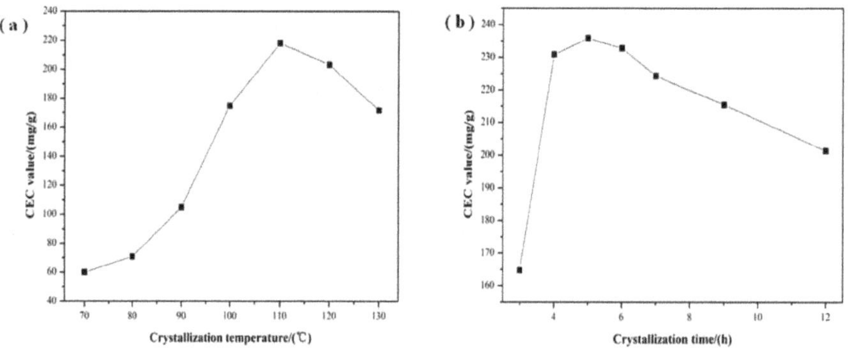

Figure 5. Effect of (**a**) crystallization temperature and (**b**) crystallization time on CEC.

3.3. Effect of Aging

Aging also played an important role in the nucleation of amorphous gel. During this stage, the structure and composition of the silica-alumina gel changed along with the aging conditions. Meanwhile, the aluminosilicate species included in the gel phase were also transformed [33]. To investigate the effect of aging conditions on the structure of the products, a batch of experiments under different aging temperatures and times were carried out by keeping the crystallization temperature and time constant at 110 °C and 5 h, respectively, and the alkalinity of the base solutions at the initial values. The XRD patterns of samples prepared at different aging temperatures and times are shown in Figure 6a,b, respectively. The XRD peaks of the prepared products at different aging temperatures exhibited similar crystallinity of zeolite X (Figure 6a) indicating that aging temperature within this range played only a minor role, if any in the formation of zeolite X. However, the synthetic product without aging exhibited extremely weak diffraction peaks of zeolite X (Figure 6b). Also, there was a sharp diffraction peak at $2\theta = 18.35°$, which could be attributed to the unreacted aluminum hydroxide. The diffraction peaks of the prepared products at different aging times exhibited the same typical features of zeolite X with high crystallization degree. However, when the aging time was up to 60 and 120 min, some obvious diffraction peaks of zeolite A appeared indicating that the aging time plays a significant role in the formation of zeolite prepared from diatomite.

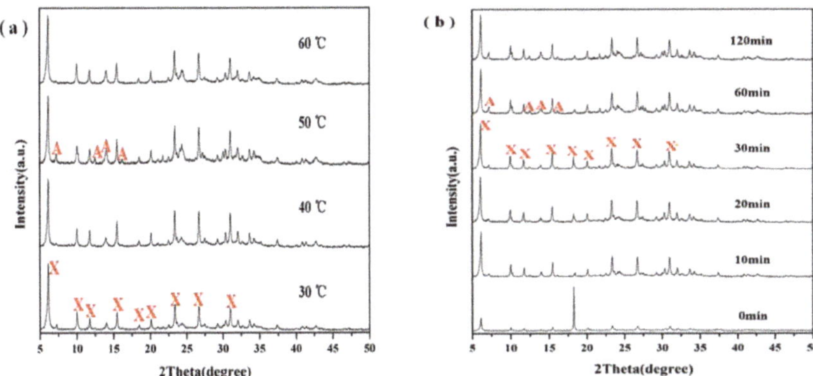

Figure 6. XRD patterns showing the effect of (**a**) aging temperature at aging time of 60 min and (**b**) aging time at aging temperature of 30 °C on zeolite formation after treatment at 110 °C and 5 h.

The CEC values of samples with different aging temperatures and times are displayed in Figure 7a,b, respectively. As shown in Figure 7a, the CEC values of samples at different aging temperatures fluctuated little. The CEC values firstly increased with the increase of aging time and then gradually decreased when the time is over 30 min (Figure 7b). Hence, on the basis of XRD and CEC analysis, samples with high crystallization degree possessed high CEC value.

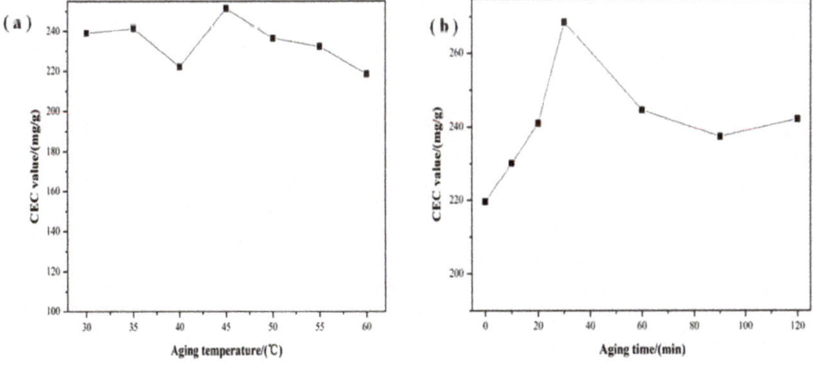

Figure 7. Effect of (**a**) aging temperature and (**b**) aging time on CEC of zeolites prepared upon hydrothermal treatment at 110 °C and 5 h.

3.4. Effect of Alkalinity

Another important parameter that controls the nucleation and crystal growth of zeolite is the alkalinity of the base solution, which includes H_2O/Na_2O and Na_2O/SiO_2 ratios [34]. Generally high alkaline concentration of the system could accelerate the dissolution of silicon and aluminum components in the precursor materials, which will shorten the induction period and nucleation time, and then speed up the crystallization rate. To investigate the effect of alkalinity on the structure of the products, a batch of experiments under different H_2O/Na_2O and Na_2O/SiO_2 ratios were carried out with the crystallization temperature and time of 110 °C and 5 h and the aging temperature and time of 30 °C and 30 min. The XRD patterns of samples with different H_2O/Na_2O and Na_2O/SiO_2 ratios are shown in Figure 8a,b, respectively. As the H_2O/Na_2O ratio increased, its crystallization degree was enhanced (Figure 8a). When the H_2O/Na_2O ratios were 40 and 45, the crystallization degree reached the maximum and there were also some zeolite A impurities. However, when the H_2O/Na_2O ratios were 45, a relatively sharp diffraction peak of aluminum hydroxide appeared (Figure 8a), which indicated that diatomite and aluminum hydroxide did not react completely. When the H_2O/Na_2O ratio was up to 50, there was only one sharp diffraction peak attributable to aluminum hydroxide. This reveals that higher H_2O/Na_2O ratios may inhibit the formation of zeolite X from diatomite, probably because low alkalinity reduces the dissolution of diatomite and results in a low conversion to zeolites. Figure 8b shows that the crystallization degree at different Na_2O/SiO_2 ratio was relatively high, which indicated that Na_2O/SiO_2 ratio within this range played minor effects in the formation of zeolite X.

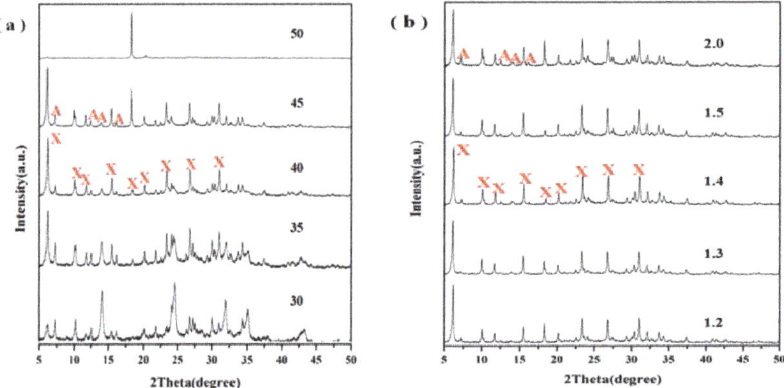

Figure 8. XRD patterns showing the effect of (**a**) H_2O/Na_2O ratio and (**b**) Na_2O/SiO_2 ratio on zeolite formation upon hydrothermal treatment at 110 °C and 5 h.

SEM images of samples with different Na_2O/SiO_2 ratios are presented in Figure 9. Figure 9 shows that all samples with Na_2O/SiO_2 ratios of 1.3, 1.4 and 1.5 exhibited high crystallinity with smooth faces. On the other hand, the sample with Na_2O/SiO_2 ratio of 1.4 exhibited better octahedral structure and uniform distribution.

Figure 9. SEM images of zeolites obtained at Na_2O/SiO_2 ratio of (**a**,**b**) 1.3, (**c**,**d**) 1.4 and (**e**,**f**) 1.5 upon hydrothermal treatment at 110 °C and 5 h.

The influence of H_2O/Na_2O and Na_2O/SiO_2 ratios on the CEC is shown in Figure 10a,b. As shown in Figure 10a, the increase of H_2O/Na_2O ratio resulted in the increase of the CEC. However, when the H_2O/Na_2O ratio was higher than 40, its CEC gradually decreased. It can be attributed to the low alkalinity, which reduces the dissolution of diatomite and results in a low conversion to zeolites. As displayed in Figure 10b, the CEC at different Na_2O/SiO_2 ratios maintained a high value and fluctuate only a little. It is obvious that higher crystallization degree led to higher CEC value. Meanwhile, the calcium ion exchange capacity of the prepared zeolite X sample is 248 mg/g, which is higher than that of commercial zeolites X (232 mg/g).

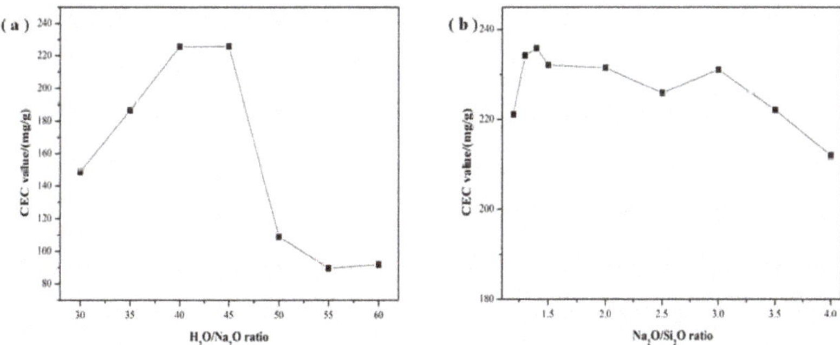

Figure 10. Effect of (**a**) H_2O/Na_2O ratio and (**b**) Na_2O/SiO_2 ratio on CEC upon hydrothermal treatment at 110 °C and 5 h.

3.5. TG-DTA and XRF Analysis

The chemical composition of the synthesized X zeolite is presented in Table 1. The synthesized X zeolite is composed of Si, Al, and Na. Meanwhile, the Si/Al ratio is 1.21, which agrees with the theoretical value of 1.13 in the raw material. In addition, the contents of Si in the raw material and the synthesized X zeolite is 3.22 and 2.88 g, respectively, which is a highly efficient use of diatomite. Thermal behavior of the obtained zeolite X sample was investigated using simultaneous TG/DTA thermoanalytical techniques. Typical TG/DTA thermograms for the prepared zeolite X sample in the temperature range of 25–900 °C are shown in Figure 11a. TG results showed that the synthesized zeolite X products lost all its moisture (22 wt %) at temperatures lower than 350 °C. In general, this kind of weight loss was due to the removal of water adsorbed on the zeolite surface and that present in the zeolite channels. Furthermore, DTA curves showed that the endothermic peaks occurred at lower temperatures (150 °C) of the synthesized zeolite X sample, which could be assigned to the loss of adsorbed water. However, the exothermic peaks at the temperature of 820 °C could be attributed to the framework collapse and crystallization of $NaAlSiO_4$ (JCPDS 52-1342), which can be attributed to the nepheline (Figure 11b). The TG-DTA analysis indicated that the synthesized X zeolite possessed excellent thermal stability.

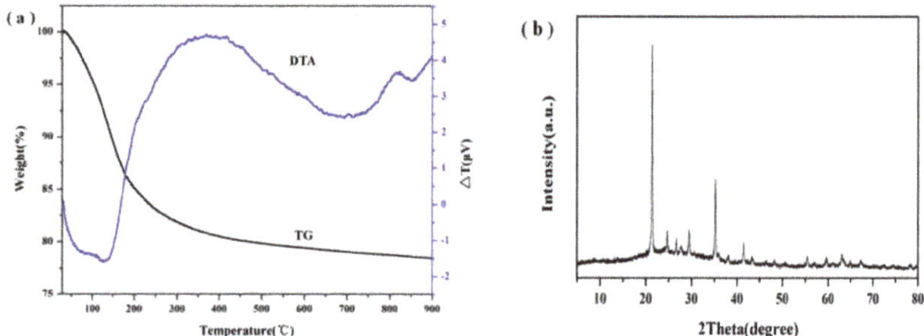

Figure 11. TG-DTA curves (**a**) of the prepared zeolite X sample at crystallization time and temperature, aging time, and temperature, H_2O/Na_2O and Na_2O/SiO_2 ratio of 5 h and 110 °C, 30 min and 30 °C, 40 and 1.4 and XRD pattern (**b**) of the prepared zeolite X sample after calcinating at 820 °C.

Table 1. The composition of the prepared zeolite X sample at crystallization time and temperature, aging time, and temperature, H_2O/Na_2O and Na_2O/SiO_2 ratio of 5 h and 110 °C, 30 min and 30 °C, 40 and 1.4.

Chemical Composition	SiO_2	Al_2O_3	Na_2O	Fe_2O_3	H_2O	$n_{(Si/Al)}$
Optimal X zeolite	35.31	24.87	15.17	0.96	22.90	1.21

3.6. N_2 Adsorption Performance

The N_2 adsorption-desorption plot at 77 K for the prepared zeolite X is presented in Figure 12. Figure 12 shows a type I isotherm with the presence of steep nitrogen uptake at very low relative pressures (p/p_0 = 0.03), which is attributed to the filling of micropores. Meanwhile, an obvious type H_4 hysteresis loop (from 0.45 to 0.99) was observed, corresponding to the filling of uniform slit-shaped intercrystal mesopores, which was ascribed to the packing of zeolite crystals [35]. Thus, the zeolite X synthesized from diatomite possessed binary meso-microporous structure. The textural properties of the prepared zeolite X are summarized in Table 2. The specific surface area and total pore volume were up to 453 m^2/g and 0.2838 cm^3/g, respectively.

Table 2. Textural properties of the zeolite X.

S_{BET} (m^2/g)	S_{micro} (m^2/g)	V_{total} (cm^3/g)	V_{micro} (cm^3/g)
453	399	0.2838	0.1866

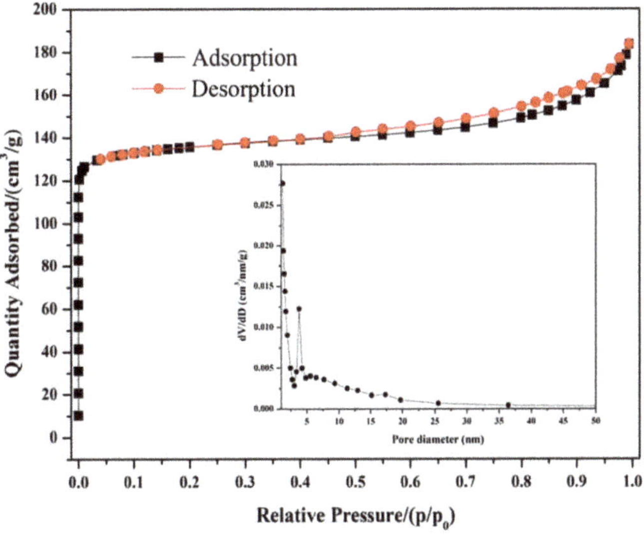

Figure 12. Nitrogen adsorption/desorption isotherms at 77 K of as synthesized zeolite X.

A comparison of the X zeolites prepared from different Si and Al sources are summarized in Table 3. The surface area of zeolite X prepared here is relatively high. Another comparison of different methods in the preparation of zeolites with diatomite as Si source is presented in Table 4. It can be seen from Table 4 that most methods need acid activation and calcination, which are not environmentally friendly and of high-cost. In addition, these conventional methods also take a longer time to achieve the transformation from minerals to zeolites. Therefore, the one-step hydrothermal method proposed in this research is more environmentally friendly and cheaper.

Table 3. Comparison of different Si and Al sources on the surface areas of synthetic zeolite X.

Zeolite	Si and Al Source	BET Surface Area (m^2/g)	Refs.
X	Fly ash	344	[36]
X	Feldspar	472	[19]
X	Bentonite	505	[19]
X	Kaolinite	591	[19]
X	Fly ash	404	[27]
X	Fly ash and sodium aluminate	397	[37]
X	Diatomite and aluminum hydroxide	453	This work

Table 4. Comparison of different methods in the preparation of zeolites with diatomite as Si source.

Zeolite	Al Source	Hydrothermal Time (h)	Synthenic Method	Refs.
Sodalite	/	/	Microwave heating methods	[38]
Y	$Al_2(SO_4)_3$	6–48	H_2SO_4 activation and hydrothermal methods	[1]
ZSM-5	$NaAlO_2$	40	Templation and hydrothermal methods	[39]
P	Aluminum hydroxide	6–24	Water-bathing methods	[40]
P	Paper sludge ash	24	Low-temperature methods	[41]
X	Aluminum hydroxide	5	Hydrothermal methods	This work

4. Conclusions

In this work, the zeolite X was obtained from diatomite via a simple hydrothermal method. In addition, the optimum preparation conditions of zeolite X were 5 h of crystallization time, 110 °C of crystallization temperature, 30 °C of aging temperature, 30 min of aging time, H_2O/Na_2O ratio of 40 and Na_2O/SiO_2 ratio of 1.4. The prepared pure zeolite X with binary meso-microporous structure possessed remarkable thermal stability, high calcium ion exchange capacity of 248 mg/g and large surface area of 453 m^2/g. Furthermore, it is shown here that the calcium ion exchange capacity of the samples is mainly determined by their crystallization degree and higher crystallization degree means higher calcium ion exchange capacity. Compared with the traditional synthesis techniques, the hydrothermal process developed here is simple, low-cost, and environmentally friendly. In addition, the high purity of zeolite X using natural low-grade diatomite as raw material may be useful for potential industrial application as catalyst support or adsorbent.

Author Contributions: Investigation, Visualization, Writing-original draft, G.Y.; Investigation, J.L.; Investigation, X.Z.; Supervision, Z.S.; Conceptualization, Supervision, S.Z.

Funding: This research is supported by the Young Elite Scientists Sponsorship Program by CAST (2017QNRC001), Yueqi Funding Scheme for Young Scholars (China university of Mining &Technology, Beijing) and the Fundamental Research Funds for the Central Universities (2015QH01 and 2010YH10).

Acknowledgments: The first author thanks the China Scholarship Council (CSC) for financial support.

Conflicts of Interest: The authors declare no conflicts of interest.

References

1. Garcia, G.; Cardenas, E.; Cabrera, S.; Hedlund, J.; Mouzon, J. Synthesis of zeolite Y from diatomite as silica source. *Microporous Mesoporous Mater.* **2016**, *219*, 29–37. [CrossRef]
2. Murad, S.; Jia, W.; Krishnamurthy, M. Molecular simulations of ion exchange in NaA zeolite membranes. *Chem. Phys. Lett.* **2003**, *369*, 402–408. [CrossRef]
3. Fan, M.; Sun, J.; Bai, S.; Panezai, H. Size effects of extraframework monovalent cations on the thermal stability and nitrogen adsorption of LSX zeolite. *Microporous Mesoporous Mater.* **2015**, *202*, 44–49. [CrossRef]
4. Uzunova, E.L.; Mikosch, H.; Hafner, J. Theoretical study of transition metal cation exchanged zeolites: Interaction with NO. *J. Mol. Struct. THEOCHEM* **2009**, *912*, 88–94. [CrossRef]
5. Tekin, R.; Bac, N. Antimicrobial behavior of ion-exchanged zeolite X containing fragrance. *Microporous Mesoporous Mater.* **2016**, *234*, 55–60. [CrossRef]
6. Doyle, A.M.; Alismaeel, Z.T.; Albayati, T.M.; Abbas, A.S. High purity FAU-type zeolite catalysts from shale rock for biodiesel production. *Fuel* **2017**, *199*, 394–402. [CrossRef]
7. Abu-Zied, B.M. Cu^{2+}-acetate exchanged X zeolites: Preparation, characterization and N_2O decomposition activity. *Microporous Mesoporous Mater.* **2011**, *139*, 59–66. [CrossRef]
8. Liu, H.; Lu, G.; Guo, Y.; Guo, Y.; Wang, J. Deactivation and regeneration of TS-1/diatomite catalyst for hydroxylation of phenol in fixed-bed reactor. *Chem. Eng. J.* **2005**, *108*, 187–192. [CrossRef]
9. Maihom, T.; Wannakao, S.; Boekfa, B.; Limtrakul, J. Density functional study of the activity of gold-supported ZSM-5 zeolites for nitrous oxide decomposition. *Chem. Phys. Lett.* **2013**, *556*, 217–224. [CrossRef]
10. Arefi Pour, A.; Sharifnia, S.; Neishabori Salehi, R.; Ghodrati, M. Adsorption separation of CO_2/CH_4 on the synthesized NaA zeolite shaped with montmorillonite clay in natural gas purification process. *J. Nat. Gas Sci. Eng.* **2016**, *36*, 630–643. [CrossRef]
11. Selishchev, D.S.; Kolinko, P.A.; Kozlov, D.V. Adsorbent as an essential participant in photocatalytic processes of water and air purification: Computer simulation study. *Appl. Catal. A Gen.* **2010**, *377*, 140–149. [CrossRef]
12. Knauth, M.; Vasenkov, S.; Kärger, J.; Fritzsche, S. Molecular dynamics study of sorbate diffusion in a simple porous membrane containing microporous nanocrystals and mesopores. *Chem. Phys. Lett.* **2009**, *479*, 95–99. [CrossRef]
13. Jin, X.; Jiang, M.Q.; Shan, X.Q.; Pei, Z.G.; Chen, Z. Adsorption of methylene blue and orange II onto unmodified and surfactant-modified zeolite. *J. Colloid Interface Sci.* **2008**, *328*, 243–247. [CrossRef] [PubMed]

14. Nibou, D.; Mekatel, H.; Amokrane, S.; Barkat, M.; Trari, M. Adsorption of Zn^{2+} ions onto NaA and NaX zeolites: Kinetic, equilibrium and thermodynamic studies. *J. Hazard. Mater.* **2010**, *173*, 637–646. [CrossRef] [PubMed]
15. Alver, E.; Metin, A.Ü. Anionic dye removal from aqueous solutions using modified zeolite: Adsorption kinetics and isotherm studies. *Chem. Eng. J.* **2012**, *200–202*, 59–67. [CrossRef]
16. Benaliouche, F.; Hidous, N.; Guerza, M.; Zouad, Y.; Boucheffa, Y. Characterization and water adsorption properties of Ag- and Zn-exchanged A zeolites. *Microporous Mesoporous Mater.* **2015**, *209*, 184–188. [CrossRef]
17. Feng, P.; Zhang, G.; Zang, K.; Li, X.; Xu, L.; Chen, X. A theoretical study on the selective adsorption behavior of dimethyl ether and carbon monoxide on H-FER zeolites. *Chem. Phys. Lett.* **2017**, *684*, 279–284. [CrossRef]
18. Belviso, C.; Cavalcante, F.; Lettino, A.; Fiore, S. A and X-type zeolites synthesised from kaolinite at low temperature. *Appl. Clay Sci.* **2013**, *80–81*, 162–168. [CrossRef]
19. Garshasbi, V.; Jahangiri, M.; Anbia, M. Equilibrium CO_2 adsorption on zeolite 13X prepared from natural clays. *Appl. Surf. Sci.* **2017**, *393*, 225–233. [CrossRef]
20. Alaba, P.A.; Sani, Y.M.; Mohammed, I.Y.; Abakr, Y.A.; Daud, W.M.A.W. Synthesis and characterization of sulfated hierarchical nanoporous faujasite zeolite for efficient transesterification of shea butter. *J. Clean. Prod.* **2017**, *142*, 1987–1993. [CrossRef]
21. Su, S.; Ma, H.; Chuan, X. Hydrothermal synthesis of zeolite A from K-feldspar and its crystallization mechanism. *Adv. Powder Technol.* **2016**, *27*, 139–144. [CrossRef]
22. Qian, T.; Li, J. Synthesis of Na-A zeolite from coal gangue with the in-situ crystallization technique. *Adv. Powder Technol.* **2015**, *26*, 98–104. [CrossRef]
23. Chen, D.; Hu, X.; Shi, L.; Cui, Q.; Wang, H.; Yao, H. Synthesis and characterization of zeolite X from lithium slag. *Appl. Clay Sci.* **2012**, *59–60*, 148–151. [CrossRef]
24. Purnomo, C.W.; Salim, C.; Hinode, H. Synthesis of pure Na–X and Na–A zeolite from bagasse fly ash. *Microporous Mesoporous Mater.* **2012**, *162*, 6–13. [CrossRef]
25. Ma, D.; Wang, Z.; Guo, M.; Zhang, M.; Liu, J. Feasible conversion of solid waste bauxite tailings into highly crystalline 4A zeolite with valuable application. *Waste Manag.* **2014**, *34*, 2365–2372. [CrossRef] [PubMed]
26. Li, C.; Zhong, H.; Wang, S.; Xue, J.; Zhang, Z. Removal of basic dye (methylene blue) from aqueous solution using zeolite synthesized from electrolytic manganese residue. *J. Ind. Eng. Chem.* **2015**, *23*, 344–352. [CrossRef]
27. Musyoka, N.M.; Ren, J.; Langmi, H.W.; North, B.C.; Mathe, M. A comparison of hydrogen storage capacity of commercial and fly ash-derived zeolite X together with their respective templated carbon derivatives. *Int. J. Hydrog. Energy* **2015**, *40*, 12705–12712. [CrossRef]
28. Babajide, O.; Musyoka, N.; Petrik, L.; Ameer, F. Novel zeolite Na-X synthesized from fly ash as a heterogeneous catalyst in biodiesel production. *Catal. Today* **2012**, *190*, 54–60. [CrossRef]
29. Volli, V.; Purkait, M.K. Selective preparation of zeolite X and A from flyash and its use as catalyst for biodiesel production. *J. Hazard. Mater.* **2015**, *297*, 101–111. [CrossRef] [PubMed]
30. Manadee, S.; Sophiphun, O.; Osakoo, N.; Supamathanon, N.; Kidkhunthod, P.; Chanlek, N.; Wittayakun, J.; Prayoonpokarach, S. Identification of potassium phase in catalysts supported on zeolite NaX and performance in transesterification of Jatropha seed oil. *Fuel Process. Technol.* **2017**, *156*, 62–67. [CrossRef]
31. Li, T.; Liu, H.; Fan, Y.; Yuan, P.; Shi, G.; Bi, X.T.; Bao, X. Synthesis of zeolite Y from natural aluminosilicate minerals for fluid catalytic cracking application. *Green Chem.* **2012**, *14*, 3255–3259. [CrossRef]
32. Magaña, S.M.; Quintana, P.; Aguilar, D.H.; Toledo, J.A.; Ángeles-Chávez, C.; Cortés, M.A.; León, L.; Freile-Pelegrín, Y.; López, T.; Sánchez, R.M.T. Antibacterial activity of montmorillonites modified with silver. *J. Mol. Catal. A Chem.* **2008**, *281*, 192–199. [CrossRef]
33. Ogura, M.; Kawazu, Y.; Takahashi, H.; Okubo, T. Aluminosilicate species in the hydrogel phase formed during the aging process for the crystallization of FAU zeolite. *Chem. Mater.* **2003**, *15*, 2661–2667. [CrossRef]
34. Liu, L.; Du, T.; Li, G.; Yang, F.; Che, S. Using one waste to tackle another: Preparation of a CO_2 capture material zeolite X from laterite residue and bauxite. *J. Hazard. Mater.* **2014**, *278*, 551–558. [CrossRef] [PubMed]
35. Ltaief, O.O.; Siffert, S.; Fourmentin, S.; Benzina, M. Synthesis of Faujasite type zeolite from low grade Tunisian clay for the removal of heavy metals from aqueous waste by batch process: Kinetic and equilibrium study. *C. R. Chim.* **2015**, *18*, 1123–1133. [CrossRef]
36. Derkowski, A.; Franus, W.; Waniak-Nowicka, H.; Czímerová, A. Textural properties vs. CEC and EGME retention of Na–X zeolite prepared from fly ash at room temperature. *Int. J. Miner. Process.* **2007**, *82*, 57–68. [CrossRef]

37. Izidoro, J.D.C.; Fungaro, D.A.; Abbott, J.E.; Wang, S. Synthesis of zeolites X and A from fly ashes for cadmium and zinc removal from aqueous solutions in single and binary ion systems. *Fuel* **2013**, *103*, 827–834. [CrossRef]
38. Zeng, S.; Wang, R.; Zhang, Z.; Qiu, S. Solventless green synthesis of sodalite zeolite using diatomite as silica source by a microwave heating technique. *Inorg. Chem. Commun.* **2016**, *70*, 168–171. [CrossRef]
39. Sanhueza, V.; Kelm, U.; Cid, R.; López-Escobar, L. Synthesis of ZSM-5 from diatomite: A case of zeolite synthesis from a natural material. *J. Chem. Technol. Biotechnol.* **2004**, *79*, 686–690. [CrossRef]
40. Du, Y.; Shi, S.; Dai, H. Water-bathing synthesis of high-surface-area zeolite P from diatomite. *Particuology* **2011**, *9*, 174–178. [CrossRef]
41. Wajima, T.; Haga, M.; Kuzawa, K.; Ishimoto, H.; Tamada, O.; Ito, K.; Nishiyama, T.; Downs, R.T.; Rakovan, J.F. Zeolite synthesis from paper sludge ash at low temperature (90 degrees C) with addition of diatomite. *J. Hazard. Mater.* **2006**, *132*, 244–252. [CrossRef] [PubMed]

Article

Mesoscale Anisotropy in Porous Media Made of Clay Minerals. A Numerical Study Constrained by Experimental Data

Thomas Dabat *, Arnaud Mazurier, Fabien Hubert, Emmanuel Tertre, Brian Grégoire, Baptiste Dazas and Eric Ferrage *

CNRS (Centre National de la Recherche Scientifique), Université de Poitiers, UMR 7285 IC2MP-Hydrasa, 5 rue Albert Turpain, Bâtiment B8, TSA 51106, 86073 Poitiers CEDEX 9, France; arnaud.mazurier@univ-poitiers.fr (A.M.); fabien.hubert@univ-poitiers.fr (F.H.); emmanuel.tertre@univ-poitiers.fr (E.T.); brian.gregoire@univ-poitiers.fr (B.G.); baptiste.dazas@univ-poitiers.fr (B.D.)

* Correspondence: thomas.dabat@univ-poitiers.fr (T.D.); eric.ferrage@univ-poitiers.fr (E.F.); Tel.: +33-549-366-395 (T.D.)

Received: 21 September 2018; Accepted: 12 October 2018; Published: 13 October 2018

Abstract: The anisotropic properties of clay-rich porous media have significant impact on the directional dependence of fluids migration in environmental and engineering sciences. This anisotropy, linked to the preferential orientation of flat anisometric clay minerals particles, is studied here on the basis of the simulation of three-dimensional packings of non-interacting disks, using a sequential deposition algorithm under a gravitational field. Simulations show that the obtained porosities fall onto a single master curve when plotted against the anisotropy value. This finding is consistent with results from sedimentation experiments using polytetrafluoroethylene (PTFE) disks and subsequent extraction of particle anisotropy through X-ray microtomography. Further geometrical analyses of computed porous media highlight that both particle orientation and particle aggregation are responsible of the evolution of porosity as a function of anisotropy. Moreover, morphological analysis of the porous media using chord length measurements shows that the anisotropy of the pore and solid networks can be correlated with particle orientation. These results indicate that computed porous media, mimicking the organization of clay minerals, can be used to shed light on the anisotropic properties of fluid transfer in clay-based materials.

Keywords: clay minerals particles; orientational anisotropy; granular systems; disk packing; X-Ray microtomography; mesoscale simulation

1. Introduction

The understanding of textural properties of natural porous media composed of clay minerals is of prime importance, owing to the key role of these minerals in retention and mobility of various resources (water, gas) and pollutants in natural environments such as soils and rocks [1–4] as well as for industrial application such as clay liners in civil engineering [5–10]. Crystal structure of clay minerals comprises one or two tetrahedral sheets (with mainly Si, Al) and one octahedral sheet (with Al, Mg, Fe, Li, etc.). At the molecular scale, exchange capacities governing reactivity of clay minerals result from negative charges created by isomorphic substitutions either in the tetrahedral and/or octahedral sheets balanced by cations [11].

Owing to their low aspect ratio (particle thickness to length ratio) coupled with large surface areas [12–15], clay minerals organization can be conveniently modelled with flat ellipsoids, disks, or elliptic disk-shaped particles [16–19]. Textural properties and associated preferential orientation of clay particles have substantial influence on the pore network morphology, such as pore size, tortuosity, and constrictivity impacting the directional dependence of fluid transfer [19–23]. In addition, the orientation of particles is inherited from initial geological conditions and may be further affected by compaction and shear forces [24–26].

Imaging techniques such as Transmission Electron Microscopy (TEM), Scanning Electron Microscopy (SEM) or tomography represent efficient methods to characterize the organization of clay minerals particles [4,20,27–30] especially for the in situ analysis of size and shape of pores. Given the multiscale nature of the clay-based materials, the spatial resolution needed to cover a wide range of porosities is still difficult to reach, especially for the smallest pore sizes. Three-dimensional (3D) particle packing simulation represents a powerful methodology for overcoming experimental limitations. As far as swelling clay minerals (i.e., smectite) are concerned, previous numerical works studied individual layers aggregation process by taking into account molecular interaction forces to derive particle formation mechanism [17,31–33]. For other types of platy-shaped clay minerals, thick particles are generally composed of a large number of firmly stacked individual layers. In such a case, the contribution from cohesive forces between particles is reduced and existing numerical works have mainly focused on the final organization of the porous medium and on the analysis of the mutual arrangements of grains in the packing [18,34,35]. In this latter case, and based on the sequential deposition algorithm from Coelho et al. [34], Ferrage et al. [18] recently showed that particle simulations considering non-interacting flat elliptic disks were able to reproduce different parameters measured on porous media composed of vermiculite particles, such as distribution of size, shape, and orientation of particle [36]. Moreover, this work also showed that both porosity and the degree of anisotropy in particle orientation have to be considered for anisometric particles.

In the present study, simulations of 3D disk packings with different anisotropy degrees are computed to better understand the geometrical properties of granular systems made of anisometric particles. While not attempting to reproduce the complexity of natural sedimentation processes, these simulations remain a good model for the analysis of preferential orientation of packed clay minerals particles. In the first section, the evolution of simulated porosity as a function of anisotropy is analyzed and validated against experimental data, provided by X-ray microtomography. Validation of simulations by experiments then allows, in the second section, the use of the computational results for a more detailed analysis of geometrical properties of these anisotropic granular systems.

2. Materials and Methods

2.1. Simulation of 3D Disk Packings

The simulated 3D disk packings were obtained using a sequential deposition algorithm of particles in a square simulation box of width w under a gravitational field [18,34]. As previously mentioned [18,34], such type of simulation does not reproduce the sedimentation process, but rather allows for generating grain packings with a wide range of anisotropic properties in particles orientations. According to this algorithm, the bottom of the simulation box (at $z = 0$) is considered rigid. Disks are introduced at the top of the box with a given initial angle. The particules then fall down until one or several contact points are detected with either the bottom of the box or the bed of settled particles. Once a contact is detected, the disk settles using the steepest descent method based on the altitude minimization of the barycenter. The disk is thus allowed to slide and swivel through a step-by-step process at a random amplitude between zero and a fixed maximum value. After each motion attempt, the position of the particle is rejected if overlapping is detected or if no gain in barycenter altitude is obtained. The present algorithm considers three cycles of 800 attempts for each particle to reach minimum altitude. Between each cycle the maximum amplitudes of translation and rotation are divided by 2 to limit the rejection of movement attempts when reaching the final settled position. Further simulation details are given in [18].

In this study, clay minerals particles are shown as flat rigid disks of diameter d = 2 μm and 0.2 μm in thickness, leading to an aspect ratio (i.e., diameter/thickness) of 0.1. This shape is in agreement with the aspect ratio of clay minerals particles experimentally measured by Reinholdt et al. [15] for vermiculite (from 0.08 to 0.10) and by Hassan et al. [37] for illite (from 0.11 to 0.12) and kaolinite (0.07). The width of the square simulation box was set at 8.5d to reach a representative elementary volume as demonstrated by Ferrage et al. [18]. Each simulation contains 5500 particles to obtain a disk packing large enough to extract a cubic sub-volume (8.5d in width). In order to obtain a wide range of organizations in the final packings, the different simulations were performed considering contrasted initial orientation of the particles (from 0 to 90°) and contrasted maximum amplitude in motion during settling (from $d/7$ to 5d and from 0 to 90° for slide and swivel movements, respectively).

The packings are analyzed in slices along the vertical z axis (thickness dz and volume w^2dz) allowing for displaying profiles of the porosity and of the orientation of particles. The porosity ε is determined by summing the disk volumes of particles whose barycenter altitude rz are found in the range $z \leq rz < z + dz$. The orientational anisotropy of particles with barycenter comprised of the slice is defined by the order parameter S [38–43]. A value of 0 describes isotropic systems whereas a value of S=1 corresponds to perfectly oriented particles. This order parameter is calculated based on the average of the second-order Legendre polynomial as follows:

$$S = \langle P_2 \rangle = \left\langle 3\cos^2\theta - 1 \right\rangle / 2, \tag{1}$$

where θ represents the angle between the normal unit vector of the particle and the z axis of the simulation box. As illustrated for the packing reported in Figure 1a, the ε and S parameters are extracted after defining z_{min} and z_{max} values corresponding to altitudes delimiting a packing with homogeneous ε and S profiles (Figure 1b). A denser and more oriented organization is systematically observed below z_{min} associated with the rigid bottom of the simulation box, leading to particles lying flat on the box surface. Furthermore, packings are more porous above z_{max} due to incomplete filling of the particle bed with no effect on the orientation of particles. Thus, all values of ε and S given hereafter were extracted in between carefully chosen z_{min} and z_{max} for each medium. An uncertainty of ±0.02 was attributed to both parameters in this study.

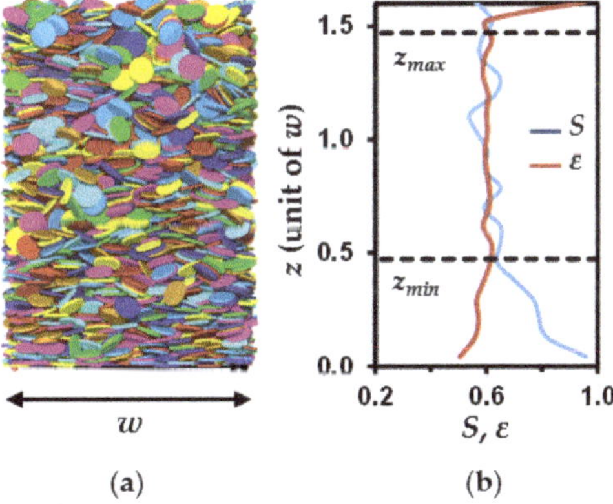

Figure 1. Analysis of porosity and anisotropy in simulated disk packings. (**a**) Three-dimensional packing and (**b**) associated porosity ε and order parameter S along the z direction. The overall porosity and S values for the packings are calculated from the nearly homogeneous part of the profiles between z_{min} and z_{max}.

2.2. Experimental Disk Packings

2.2.1. Experimental Setup and Preparation of Disk Packings

Experiments were performed using Polytetrafluoroethylene (PTFE) disks of aspect ratio = 0.1 to allow a detailed comparison between simulated and experimental data. The disks were designed with a diameter of 1 cm (i.e., 1 mm thickness) as a good compromise between the machining precision on the dimension of the PTFE disks and the resolution on X-ray microtomography images used for ε and S measurements. About 10,000 disks were allowed to sediment in a poly(methyl methacrylate) (PMMA) cylindrical column of 12 cm diameter (Figure 2a). This sedimentation column is composed of two volumes separated by a drilled grill, both filled by the same fluid. While the top part contains the disks, the bottom part allows for gently evacuating the fluids through a discharge valve without perturbing the particle bed.

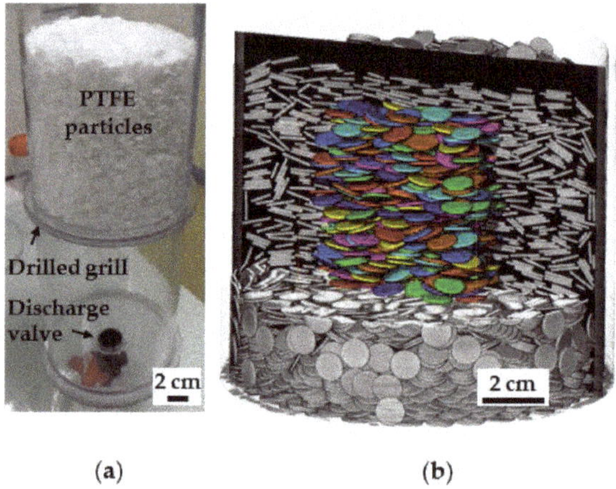

Figure 2. Experimental setup for packing of polytetrafluoroethylene (PTFE) disks. (**a**) Cylindrical column for sedimentation of 10,000 disks in different fluids. Disks are settled in the volume located above a flat drilled grill while evacuation of fluids is achieved through a discharge valve located at the bottom of the column; (**b**) X-ray microtomography analysis and extraction of individual particle orientation in a sub-volume of the experimental packing.

To obtain various overall packing organizations, different fluids with various density values were considered in order to adjust the settling rate of particles. A total of 5 experimental disk packings, hereafter referred to as DP1 to DP5, were obtained according to the setting parameters reported in Table 1. DP1 to DP4 differ by the density of the fluids considered from almost 0 (air) to 2.15 (Na-polytungstate) while DP5 has a different particles drop-off strategy. Na-polytungstate is highly soluble in water and thus allows for preparing a controlled density solution depending on its concentration [44,45]. A density of 2.15 for the Na-polytungstate solution represents the maximum value allowing sedimentation of PTFE disks (density of 2.16). Excepted for DP1 where particles were gently deposited on top of the existing bed, for all other DPs, the column was first filled with the selected fluid and the particles were let to settle from a high altitude (typically > 20 cm). From DP1 to DP4 all particles were dropped individually to mimic the one-by-one deposition algorithm with a high value for initial angle. DP5 was prepared as DP4 in a 2.15 density fluid but the particles were dropped all at once, in an attempt to obtain a less anisotropic organization (Table 1). After sedimentation of PTFE disks, liquids were eliminated through the discharge valve and DP3 to DP5 were rinsed 7 times using distilled water to remove Na-polytungstate. Finally, the DPs were dried at room conditions before X-ray microtomographic analysis.

2.2.2. X-Ray Microtomography Analyses

The X-ray microtomographic acquisitions were performed on an EasyTom XL Duo from RX-solutions (Chavanod, France). A sealed microfocus X-ray source L 12161-07 (Hamamatsu Photonics, Hamamatsu, Japan) was used, coupled to a Varian PaxScan 2520DX detector flat panel with amorphous silicon and a CsI conversion screen; 1920 × 1536 pixel matrix; pixel pitch of 127 μm; 16 bits of dynamic (Varian Medical Systems, Palo Alto, CA, USA). The entire samples were scanned with a spatial resolution of 73.36 μm in a helicoid mode (4320 projections in three turns) in order to reduce cone beam. Acquisition parameters were set at 150 kV for tube voltage and 420 μA for tube current. A setting of 12.5 frames per second was used with an averaging of 20 frames per projections. Filtration of the beam was performed using a 1 mm Al filter. The scanning was realized with the large focus mode (i.e., a nominal focal spot size of 50 μm) for a source-to-detector distance of 419 mm and a source-to-object distance of 242 mm. Data reconstruction was achieved using the XAct software v1.1 (RX-solutions, Chavanod, France) with a classical filtered back projection algorithm [46] to correct from beam hardening and ring artefacts. The segmentation of the solid vs. void, porosity, and orientation measurements, as well as visual rendering, were performed on a sub-volume (Figure 2b) using Avizo Software v.9.2 (FEI, Hillsboro, OR, USA) with an adapted image processing methodology mostly based on mathematical morphology tools [47,48]. This latter processing methodology is detailed in the Supplementary Material and accounts for a sequence of data treatments (Figure S1), from the raw image to the segmentation of individual disk particles (Figure 2b).

Table 1. Settling conditions (density of fluid and particle drop-off procedure chosen) for disk packing experiments and associated extraction of porosity ε and order parameter S (uncertainty of ±0.02).

Sample	DP1	DP2	DP3	DP4	DP5
Fluid	Air	Water	Na-polytungstate	Na-polytungstate	Na polytungstate
Density	10^{-3}	1.00	2.10	2.15	2.15
Drop-off	Individual	Individual	Individual	Individual	All at once
ε	0.48	0.47	0.49	0.54	0.51
S	0.96	0.97	0.91	0.90	0.84

3. Results and Discussion

3.1. Evolution of Simulated and Experimental Porosity with Packing Anisotropy

About 135 disk packings were simulated to cover the whole range of anisotropy, i.e., from S~0 to S~1 (Figure 3a). This was achieved by tuning the degree of freedom in particle motions, based on the input values given to initial angles and maximum amplitudes in swivel or slide motions for particles (see Section 2.1 for the description of the algorithm or [18] for a systematic analysis on the influence of individual parameters on the final packing). As an illustration, low initial angles combined with high maximum amplitudes (i.e., high degree of freedom in particle motions) allow the particle to explore a large number of positions and orientational configurations to minimize its barycenter altitude. This leads to high S and low ε values whereas, on the contrary, more porous and isotropic packings are obtained when limiting motion amplitudes with high initial angles. As shown in Figure 3a and irrespective of the degrees of freedom given to the particles, all simulated data exhibit the same tendency of decreasing porosity coupled with an increase of the orientational anisotropy (Figure 3a). A change of trend for ε values at $S > 0.9$ is also noticed on this Figure 3a. Although marginal, the dispersion of ε values for a given anisotropy S typically results from the role played by individual input parameters on the final packing configurations. As an example, for low degree of freedom in particle motions, the disks can be trapped in local minima leading to the formation of arches structures and to large porosities [49–51]. The whole set of data points can be used to confirm the presence of a master curve correlating ε and S when using this sequential deposition algorithm [18]. This master curve is highlighted in Figure 3b when selecting, from the 135 samples, 15 packings with the lowest ε value for a given anisotropy S. Input parameters used to obtain these reference packings are reported in Table 2.

For experimental packings, the extracted ε and S values are reported in Table 1. The high degree of anisotropy obtained for all packings ($0.84 < S < 0.97$) can be assigned to two principal effects. First, despite the efforts to control the contrast in density between the fluid and the PTFE disks, settling particles were observed to exhibit a low angle value when entering in contact with the existing bed of disks at the bottom of the fluid column. Second, once in contact, the settling PTFE disks were noticed to easily slide on top of other particles, likely associated with the smooth surface of PTFE disks. As pointed out from the simulated packings (Table 2), low degrees of anisotropy (typically for $S < 0.80$) are strongly linked to the initial angle of deposition (typically above 80°) and to the limited extent of slide amplitude ($\leq d$). These unreachable conditions in our experiments likely explain the limited range obtained for S values (Table 1). Despite this limited extent of anisotropy variation, a negative evolution with ε is noticed when increasing S (Figure 3b). Furthermore, an overall good agreement is found when comparing experimental and simulated ε values for a given S. These relatively high ε values obtained for both experimental and simulated packings are also consistent with experimental data reported for platy-shaped or tubular clay minerals [43,52–55]. Note the tendency to have slightly lower experimental ε values for porous media prepared in Na-polytungstate fluids (DP3 to DP5) compared to simulations, however. The slightly denser packings obtained experimentally can be first assigned to the fact that PTFE disks were observed to form aggregates through face-to-face contact once immerged in the fluid (DP5) and/or when settling on the bed of particles (DP3 to DP5). As illustrated for DP5, the difference between the experimental ε value and that expected to lie on the master curve issued from packing simulations, is of ~0.05 (Figure 3b). This aggregation effect is similar to the sedimentation of disks with higher aspect ratios. As shown by Ferrage et al. [18], an increase of aspect ratio from 0.1 to 0.2 leads to a decrease of ~0.15 in ε value for S~0.8. Likewise, the work of Ebrahimi et al. [17], using a different algorithm clearly evidenced a significant decrease of porosity coupled with an increase of aggregation. Moreover, X-ray tomographic imagery is a blurry approximation of a true material configuration [56]. Indeed, the structures detectability in tomographic images mainly depends on the inherent resolution limitations of the system and on the partial volume effect (i.e., the value of a voxel covering multiple materials correspond to an average attenuation). Thereby, the smallest pores, such as those located near face-to-face contact of particles in very anisotropic media, are extremely difficult to segment. As a result, a marginal fraction of the smallest pores is omitted and thus slightly lowers the ε value. Despite the marginal differences between experimental and simulated ε values (Figure 3b), the very good agreement obtained for $S \geq 0.84$ tends to validate the sequential deposition algorithm used here.

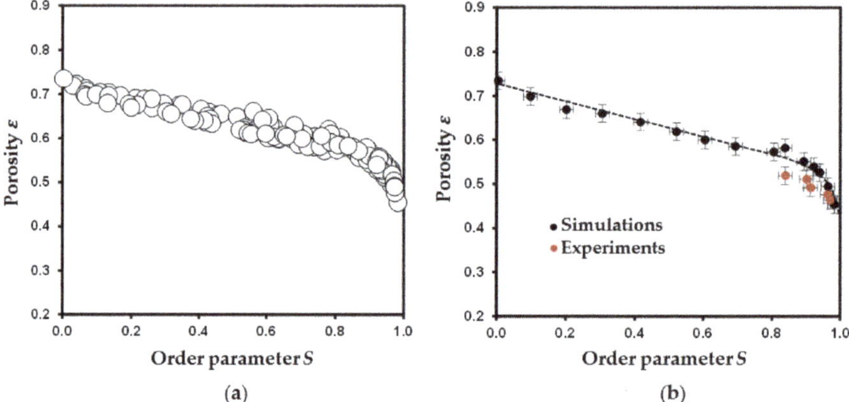

Figure 3. Evolution of porosity ε as a function of order parameter S for simulated disk packings. (**a**) The 135 disk packings obtained with different degree of freedom in particle motions. (**b**) Selection of 15 disk packings (solid circles) and comparison with experimental porous media (solid red circles). The doted curve highlighting the ε vs. S master curve is plotted as guideline.

3.2. Comparison between Simulated and Experimental Orientation Distribution Functions

The comparison between experimental and simulated data performed above is limited to ε and S values, which are bulk parameters of the porous media. For instance, different particle organizations can in principle lead to the same macroscopic S values. Accordingly, the orientation distribution function (ODF) of particles for experimental and simulated are analyzed and compared below. Because the frame of the simulation box merges with that of the orientation tensor [18], the ODF for disk-shaped particles can be defined here by the function $f(\theta, \phi)$, where θ is angle between the normal of the disk and the z axis of the simulation box and ϕ the polar azimuthal angle on the (xy) plane of the simulation box. For uniaxial systems (cylindrical symmetry) such as those investigated here the ODF is completely described by the function $f(\theta)$ defined as:

$$f(\theta) \geq 0 \tag{2}$$

and

$$\int_0^{\pi} f(\theta) \sin\theta d\theta = 1 \tag{3}$$

where $\sin\theta$ accounts for the integration over all azimuthal ϕ angles for the solid angle correction [43,57,58].

Table 2. Algorithm parameters used to generate the selected 15 simulated disk packings. For each medium, all particles are given an initial angle (in degree) and maximum amplitudes motions to swivel (in degree) and to slide (relative to particle diameter d) during the settling process. The parameters S and ε stand for the order parameter and the porosity, respectively (uncertainty of ± 0.02). $N_{part.}$ and $f_{part.}$ represent the mean number of particles and the fraction of particles in aggregates, respectively.

Initial Angle	Max. Swivel Amplitude (°)	Max. Slide Amplitude (μm)	S	ε	$N_{part.}$	$f_{part.}$
82	11	$d/6.4$	0.00	0.73	2.28	0.34
87	22	$d/7$	0.10	0.70	2.33	0.42
87	25	$d/2$	0.20	0.67	2.42	0.47
85	40	$d/7$	0.31	0.66	2.48	0.51
89	50	$d/7$	0.42	0.64	2.55	0.56
82	60	$d/2$	0.52	0.62	2.69	0.59
85	85	$d/3$	0.60	0.60	2.80	0.65
85	80	d	0.69	0.59	2.85	0.63
60	60	d	0.81	0.57	2.91	0.69
47	28	$d/3$	0.84	0.58	2.90	0.67
40	40	d	0.89	0.55	3.06	0.73
25	45	$d/2$	0.92	0.54	3.09	0.74
18	55	d	0.94	0.53	3.16	0.76
10	70	$4d$	0.96	0.49	3.63	0.83
0	80	$5d$	0.98	0.45	3.91	0.87

Figure 4 reports $f(\theta)sin\ \theta$ functions for different S values. For comparison with experimental systems (S = 0.84; 0.90, and 0.96), simulated porous media with similar S values are selected from Table 2 (S = 0.84; 0.89, and 0.96). For an isotropic packing (S = 0), $f(\theta)sin\ \theta$ follows a sine behavior. In the case where S tends to 1, $f(\theta)sin\ \theta$ function displays a Dirac-shaped distribution. For similar anisotropy value, it can be shown that the experimental and simulated data show very good agreement, despite the limited number of extracted particles from the experiments (between 671 and 920 particles for experimental media vs. between 3270 and 3955 for simulated packings). This agreement shows that the sequential deposition algorithm used here provides, not only a satisfying reproduction of bulk properties of the experimental porous media, but also, accounts for the details of local particle organization.

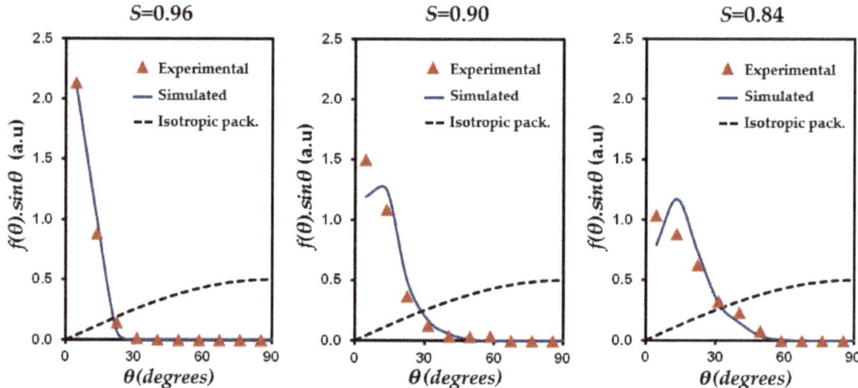

Figure 4. Comparison between simulated and experimental distributions of particle orientation $f(\theta)\sin\theta$ with θ the angle between the normal unit vector of the particle and the z axis of the simulation box. The selected experimental packings correspond to DP1, DP4, and DP5 (with S = 0.96, 0.90, and 0.84, respectively, Table 1), whereas simulated packings were chosen in Table 2 for their similar S values (with S = 0.96, 0.89, and 0.84).

3.3. Evolution of Geometrical Properties of Simulated Porous Media with Anisotropy

In the following, the 15 simulated porous media (Table 2) are used to get additional information on the evolution of the geometrical properties of the pore network and the solid phase with S parameter. This morphological analysis, based bellow on chord length distribution and particle aggregation, is expected to provide a comprehensive understanding of the underlying mechanism at the origin of the observed master curve of ε vs. S and in particular to the change of porosity trend above S~0.9 (Figure 3b). Note that all attempts to apply the same morphological analysis for experimental systems were unsuccessful. This was assigned to the difficulty to detect the smallest pores at the interface between two particles, thus leading to erroneous chord length distribution analyses.

A chord length analysis provides a morphological description of the solid-pore interface [18,59–62]. Chords are line segments at the pore-solid interface, for any direction r, lying entirely in the pore or solid phase. For the pore phase, the pore chord distribution $f_p(r)$ is defined such as $f_p(r)dr$ is the probability of finding a chord in the pore phase of a length between r and $r + dr$ and by:

$$\int_0^\infty f_p(r)dr = 1. \tag{4}$$

Based on both the pore chord distribution $f_p(r)$ and the solid chord distribution $f_s(r)$, the mean chord length for both pore and solid phases along r ($l_{p,\bar{r}}$ and $l_{s,\bar{r}}$, respectively) are determined from the first momentum of the distribution function as follows:

$$l_{p,\bar{r}} = \int_0^\infty r\, f_p(r)dr, \tag{5}$$

$$l_{s,\bar{r}} = \int_0^\infty r\, f_s(r)dr, \tag{6}$$

where $l_{p,\bar{r}}$ and $l_{s,\bar{r}}$ can be considered to account for the mean sizes of pores and the solid phase along the same direction, respectively. Calculations of pore and solid chord length distributions are performed using 3D voxelized image (512^3 resolution) for the 15 porous media reported in Table 2.

The mean chord lengths for the solid phase of the 15 simulated porous media (Table 2) is reported for the three main directions x, y, and z as a function of the order parameter S in Figure 5a. Mean chords lengths along the x and y directions ($l_{s,\bar{x}}$ and $l_{s,\bar{y}}$) are identical, thus confirming the transverse isotropy in simulated packings.

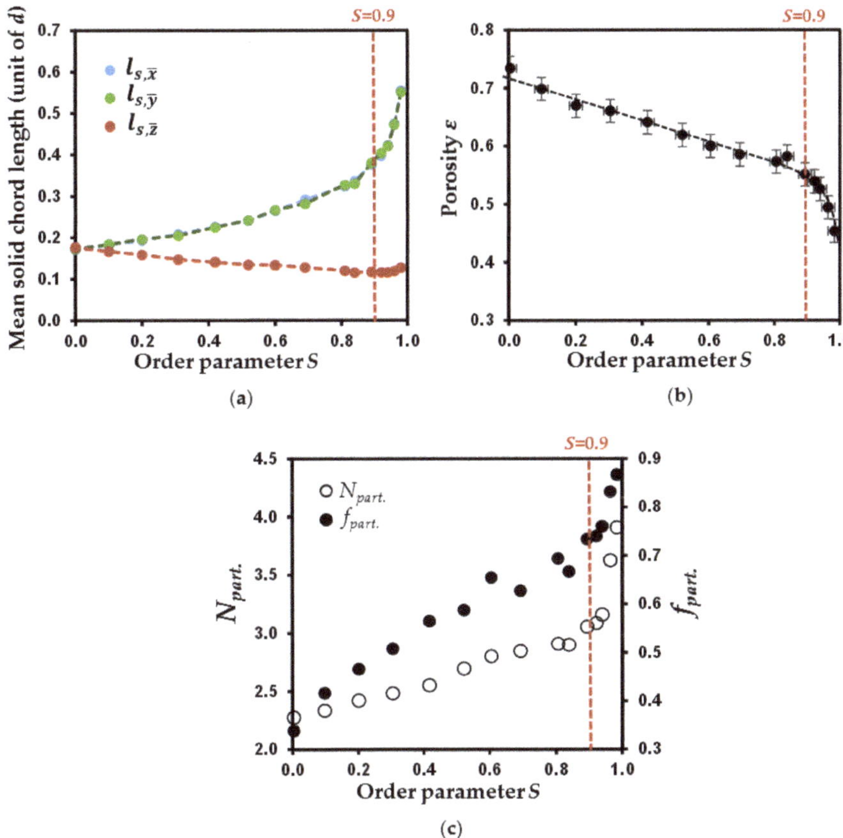

Figure 5. Evolution of geometrical parameters of the simulated porous media as a function of order parameter S. Mean chord length for the solid phase in the three main directions x, y, and z (i.e., $l_{s,\bar{x}}$, $l_{s,\bar{y}}$, and $l_{s,\bar{z}}$, respectively) (**a**), porosity ε (**b**), and mean number of particles $N_{part.}$ and fraction of particles $f_{part.}$ in aggregates (**c**). Vertical dotted line is a guideline for the change noticed at $S > 0.9$.

For $S = 0$, the isotropic organization of the packing is corroborated by the equal values for $l_{s,x}$, $l_{s,y}$ and $l_{s,\bar{z}}$. Figure 5a also reveals that the increase in particle anisotropy leads to a progressive decrease of $l_{s,\bar{z}}$ values and a gradual increase of $l_{s,\bar{x}}$ and $l_{s,\bar{y}}$ values. This evolution is fully consistent with the gradual alignment of particles axes and dimensions with that of the simulation box. Indeed $l_{s,\bar{z}}$ dimension decreases toward a value close to $d/10$ (with d the disk diameter), i.e., the thickness of individual disk. Similarly, both $l_{s,\bar{x}}$ and $l_{s,\bar{y}}$ dimensions increase to approach the theoretical mean transverse dimension of the disk at ~0.63d. Moreover, note that $l_{s,\bar{z}}$ values slightly increase with S when $S > 0.9$ (Figure 5a). This behavior seems to correlate with the abrupt decrease of porosity with S increasing for $S > 0.9$ (Figure 5b). Such concomitant increase in $l_{s,\bar{z}}$ dimensions and decrease in ε values for $S > 0.9$ could potentially be interpreted by aggregation of particles, defined here by an increased number of face-to-face contact between their flat surfaces leading to an increase of their apparent thickness. The analysis of particle aggregation is thus performed in order to assess its potential influence on the change of porosity for high S values. This analysis is performed in a similar fashion as done by Ebrahimi et al. [17,31,32] for the characterization of aggregation of individual nanometer-sized clay layers. The consideration that two particles belong to the same aggregate is based here on three criterions. The first criterion is that the scalar product between the normal vectors associated to the two particles should be larger than 0.95. This ensures the pseudo-parallelism of the two particles surfaces. The second criterion is that the segment formed by the projection of the barycenter position of the first particle onto the surface of the second disk should be shorter than the thickness of the disk. Accordingly, the two particles should first neighbor, while allowing a certain degree of non-parallelism. The third and last criterion is that the distance segment between the two barycenter of the two particles, projected to the surface of one particle, should be shorter than the disk radius. This last criterion allows for a relative lateral displacement between two neighboring particles. While the first criterion is similar to Ebrahimi et al. [17,31,32], these latter authors considered only a second criterion related to the distance between the two particles barycenter. In the case of aggregation of individual clay layers, the consideration of these two criterions leads to the formation of cylindrically stacks of particles with limited lateral misfits, which is fully consistent with the involved cohesive forces [17,31,32]. In the case of aggregation of clay minerals particles constituted by large number of individual layers, contribution from cohesive forces between particles is lowered (or neglected in the present numerical approach), and the aggregates display larger degree of lateral displacements as repeatedly observed here in X-ray microtomographic images.

The evolution of particle aggregation with S parameter, calculated according to the aforementioned methodology for the 15 simulated porous media, is illustrated in Figure 5c. The mean number of particles in the stack, $N_{part.}$ is calculated from the histograms of stack sizes (Table 2). These histograms are found to systematically follow a lognormal-shaped thickness distribution, consistent with different numerical or experimental studies [15,31,32]. In addition, the relative fraction of particles involved in a stack, $f_{part.}$ is also indicated (Figure 5c; Table 2). As seen in Figure 5c and illustrated for selected porous media in Figure 6, both $N_{part.}$ and $f_{part.}$ parameters display a rather monotonic evolution for $S < 0.9$ and then a significant rise for very anisotropic media. This behavior fully supports the hypothesis that aggregation is of the origin of the abrupt decrease of porosity with an increase of S when $S > 0.9$ (Figure 5b).

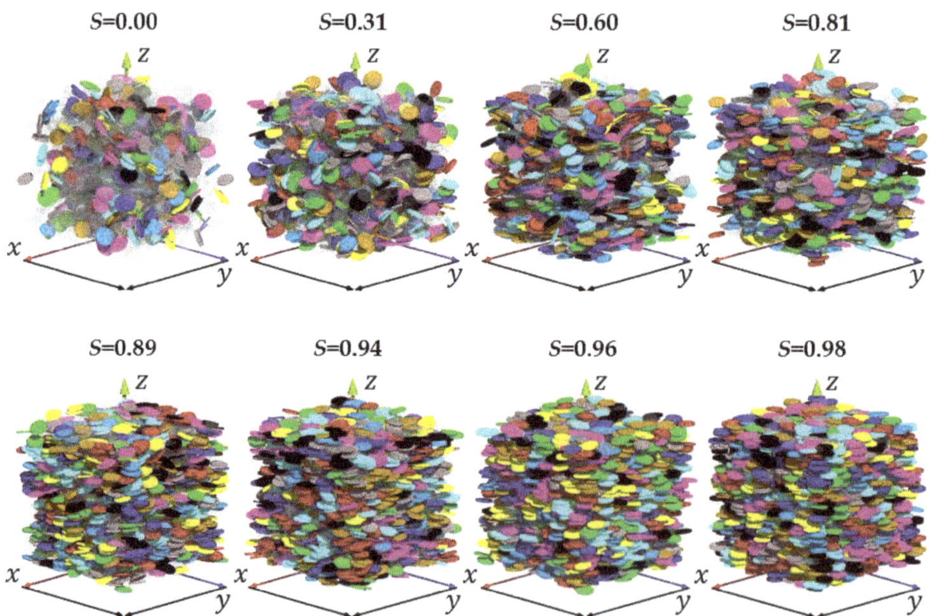

Figure 6. Illustration of particle aggregation for selected simulated packings. A given colour is assigned to all particles from the same aggregate. Translucent particles correspond to disks not involved in aggregates.

If particle orientation and aggregation can be used to interpret the evolution of ε with S, analysis of chord distribution can also be used to derive quantitative indicators of the anisotropy of the solid and pore phases. This can be done by calculating ratios between mean chord length along the z and xy directions as:

$$R_p = l_{p,\overline{z}}/l_{p,\overline{xy}}, \tag{7}$$

$$R_s = l_{s,\overline{z}}/l_{s,\overline{xy}}. \tag{8}$$

Evolution of R_p and R_s parameters with S parameter reported in Figure 7 evidences that anisotropy in the pore or solid phase are fairly similar. These anisotropy indicators range from nearly 1 when S = 0 to ~0.2 for the most anisotropic medium and show a surprising linear correlation with S (Figure 7). This finding suggests that extraction of S parameter, easily derived from experimental techniques such as diffraction methods [36,41,43,63–65], can potentially be used to extract information regarding anisotropy in the pore network, this latter being difficult to obtain experimentally.

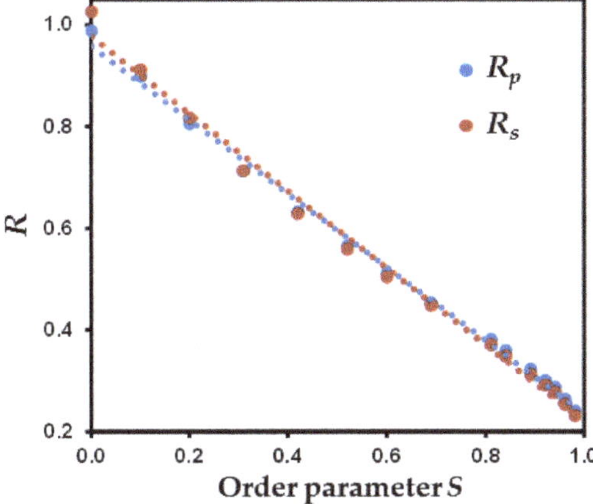

Figure 7. Evolution of ratios between mean chord length between the z and xy directions for both the solid and pore phases (R_s and R_p, respectively) as a function the order parameter S.

4. Conclusions

Simulation of 3D disk packings is an efficient approach to deepen our understanding of the role played by anisotropy in orientation of flat particles on the geometrical properties of the whole porous medium. This is particularly relevant in the case of compacted anisometric particles which can display a wide range of S values [18,41,43].

In this study, non-interacting disk packing simulations confirmed the close relationship between porosity and anisotropy through the ε vs. S master curve. Although limited to very anisotropic systems (for S > 0.8), experiments also validated the obtained correlation. Additional analyses of evolution of geometrical parameters with anisotropy of the porous media demonstrated that the significant decrease of porosity for S > 0.9 was associated to both particle orientation and particle aggregation. Interestingly this aggregation is noticed here even though no interaction forces are considered in between particles during the simulation of settling process.

Morphological analyses of the porous media through chord length measurements show that R_p and R_s parameters fall onto a linear correlation with the order parameter S. This relation is particularly relevant for correlating the orientational properties of particles easily accessible experimentally with anisotropy in the pore network. In this regard, the logical perspective of this work is to analyze the influence of S parameter on the anisotropy of diffusional properties of water in the porous medium.

Supplementary Materials: This Supplementary Material details the processing methodology used in this article to segment individual disk particles from raw X-ray microtomographic images. It is available online at http://www.mdpi.com/1996-1944/11/10/1972/s1, Figure S1: Processing methodology for porosity measurement and extraction of individual particle orientation (see text for details): (**a**) Non-local mean image (central vertical section in the sub-volume), (**b**) gradient image, (**c**) segmentation result of porosity vs. disks, (**d**) erosion of c, (**e**) distance map of d, (**f**) segmentation of e, (**g**) subtraction off from d, (**h**) erosion with a diskoid structuring element, (**i**) labelled markers, (**j**) watershed transformation of the gradient image b using the markers in i, (**k**) final result after manual correction of markers and quantitative filtration, (**l**) 3D rendering of the final result.

Author Contributions: Conceptualization, T.D., F.H., E.T. and E.F.; Data curation, T.D., A.M., F.H., B.G. and B.D.; Funding acquisition, E.F.; Methodology, A.M. and F.H.; Software, E.F.; Supervision, E.F.; Writing—original draft, T.D., A.M. and E.F.; Writing—review and editing, T.D., A.M., B.G., B.D. and E.F.

Funding: The results presented are part of the Ph.D. thesis of Thomas Dabat granted by "Région Nouvelle-Aquitaine", University of Poitiers, France. The authors are grateful to CNRS interdisciplinary "défi Needs", through its "MiPor" program (Project TRANSREAC) for providing financial support for this study. Additional financial support from the European Union (ERDF) and "Région Nouvelle Aquitaine" is also acknowledged.

Acknowledgments: Claude Veit (IC2MP, Poitiers, France) is thanked for the design and conception of the PMMA columns for sedimentation experiments. The manuscript was improved by the constructive comments of three anonymous reviewers.

Conflicts of Interest: The authors declare no conflict of interest.

References

1. Altmann, S.; Tournassat, C.; Goutelard, F.; Parneix, J.C.; Gimmi, T.; Maes, N. Diffusion-driven transport in clayrock formations. *Appl. Geochem.* **2012**, *27*, 463–478. [CrossRef]
2. Aristilde, L.; Galdi, S.M.; Kelch, S.E.; Aoki, T.G. Sugar-influenced water diffusion, interaction, and retention in clay interlayer nanopores probed by theoretical simulations and experimental spectroscopies. *Adv. Water Resour.* **2017**, *106*, 24–38. [CrossRef]
3. Gu, X.; Sun, J.; Evans, L.J. The development of a multi-surface soil speciation model for Cd (II) and Pb (II): Comparison of two approaches for metal adsorption to clay fractions. *Appl. Geochem.* **2014**, *47*, 99–108. [CrossRef]
4. Leu, L.; Georgiadis, A.; Blunt, M.; Busch, A.; Bertier, P.; Schweinar, K.; Liebi, M.; Menzel, A.; Ott, H. Multiscale description of shale pore systems by scanning SAXS and WAXS microscopy. *Energy Fuels* **2016**, *30*, 10282–10297.
5. Czurda, K.A.; Wagner, J.F. Cation transport and retardation processes in view of the toxic waste deposition problem in clay rocks and clay liner encapsulation. *Eng. Geol.* **1991**, *30*, 103–113.
6. Harvey, C.; Lagaly, G. Industrial applications. In *Developments in Clay Science*; Elsevier: Amsterdam, The Netherlands, 2013; pp. 451–490, ISBN 978-0-08-099364-5.
7. Lee, J.M.; Shackelford, C.D.; Benson, C.H.; Jo, H.Y.; Edil, T.B. Correlating index properties and hydraulic conductivity of geosynthetic clay liners. *J. Geotech. Geoenviron. Eng.* **2005**, *131*, 1319–1329. [CrossRef]
8. Madsen, F. Clay mineralogical investigations related to nuclear waste disposal. *Clay Miner.* **1998**, *33*, 109–129.
9. Ortiz, L.; Volckaert, G.; Mallants, D. Gas generation and migration in Boom Clay, a potential host rock formation for nuclear waste storage. *Eng. Geol.* **2002**, *64*, 287–296. [CrossRef]
10. Roehl, K.E.; Czurda, K. Diffusion and solid speciation of Cd and Pb in clay liners. *Appl. Clay Sci.* **1998**, *12*, 387–402. [CrossRef]
11. Brigatti, M.; Galan, E.; Theng, B. Structure and mineralogy of clay minerals. In *Developments in Clay Science*; Elsevier: Amsterdam, The Netherlands, 2013; pp. 21–81, ISBN 978-0-08-099364-5.
12. Dias, N.M.; Gonçalves, D.; Leite, W.C.; Brinatti, A.M.; Saab, S.C.; Pires, L.F. Morphological characterization of soil clay fraction in nanometric scale. *Powder Technol.* **2013**, *241*, 36–42. [CrossRef]
13. Gélinas, V.; Vidal, D. Determination of particle shape distribution of clay using an automated AFM image analysis method. *Powder Technol.* **2010**, *203*, 254–264. [CrossRef]
14. Hubert, F.; Caner, L.; Meunier, A.; Ferrage, E. Unraveling complex <2 μm clay mineralogy from soils using X-ray diffraction profile modeling on particle-size sub-fractions: Implications for soil pedogenesis and reactivity. *Am. Mineral.* **2012**, *97*, 384–398.

15. Reinholdt, M.X.; Hubert, F.; Faurel, M.; Tertre, E.; Razafitianamaharavo, A.; Francius, G.; Prêt, D.; Petit, S.; Béré, E.; Pelletier, M. Morphological properties of vermiculite particles in size-selected fractions obtained by sonication. *Appl. Clay Sci.* **2013**, *77*, 18–32. [CrossRef]
16. Bacle, P.; Dufrêche, J.F.; Rotenberg, B.; Bourg, I.C.; Marry, V. Modeling the transport of water and ionic tracers in a micrometric clay sample. *Appl. Clay Sci.* **2016**, *123*, 18–28. [CrossRef]
17. Ebrahimi, D.; Pellenq, R.J.M.; Whittle, A.J. Mesoscale simulation of clay aggregate formation and mechanical properties. *Granul. Matter* **2016**, *18*, 49. [CrossRef]
18. Ferrage, E.; Hubert, F.; Tertre, E.; Delville, A.; Michot, L.J.; Levitz, P. Modeling the arrangement of particles in natural swelling-clay porous media using three-dimensional packing of elliptic disks. *Phys. Rev. E* **2015**, *91*, 062210. [CrossRef] [PubMed]
19. Tyagi, M.; Gimmi, T.; Churakov, S.V. Multi-scale micro-structure generation strategy for up-scaling transport in clays. *Adv. Water Resour.* **2013**, *59*, 181–195. [CrossRef]
20. Backeberg, N.R.; Iacoviello, F.; Rittner, M.; Mitchell, T.M.; Jones, A.P.; Day, R.; Wheeler, J.; Shearing, P.R.; Vermeesch, P.; Striolo, A. Quantifying the anisotropy and tortuosity of permeable pathways in clay-rich mudstones using models based on X-ray tomography. *Sci. Rep.* **2017**, *7*, 14838. [CrossRef] [PubMed]
21. Greenkorn, R.; Johnson, C.; Shallenberger, L. Directional permeability of heterogeneous anisotropic porous media. *Soc. Pet. Eng. J.* **1964**, *4*, 124–132. [CrossRef]
22. Jacops, E.; Aertsens, M.; Maes, N.; Bruggeman, C.; Krooss, B.; Amann-Hildenbrand, A.; Swennen, R.; Littke, R. Interplay of molecular size and pore network geometry on the diffusion of dissolved gases and HTO in Boom Clay. *Appl. Geochem.* **2017**, *76*, 182–195. [CrossRef]
23. Van Loon, L.R.; Soler, J.M.; Müller, W.; Bradbury, M.H. Anisotropic diffusion in layered argillaceous rocks: A case study with Opalinus Clay. *Environ. Sci. Technol.* **2004**, *38*, 5721–5728. [CrossRef] [PubMed]
24. Aplin, A.C.; Matenaar, I.F.; McCarty, D.K.; van Der Pluijm, B.A. Influence of mechanical compaction and clay mineral diagenesis on the microfabric and pore-scale properties of deep-water Gulf of Mexico mudstones. *Clays Clay Miner.* **2006**, *54*, 500–514. [CrossRef]
25. Vasseur, G.; Djeran-Maigre, I.; Grunberger, D.; Rousset, G.; Tessier, D.; Velde, B. Evolution of structural and physical parameters of clays during experimental compaction. *Mar. Pet. Geol.* **1995**, *12*, 941–954. [CrossRef]
26. Zhang, S.; Tullis, T.E.; Scruggs, V.J. Permeability anisotropy and pressure dependency of permeability in experimentally sheared gouge materials. *J. Struct. Geol.* **1999**, *21*, 795–806. [CrossRef]
27. Chen, X.; Verma, R.; Nicolas Espinoza, D.; Prodanović, M. Pore-scale determination of gas relative permeability in hydrate-bearing sediments using X-ray computed micro-tomography and lattice boltzmann method. *Water Resour. Res.* **2018**, *54*, 600–608. [CrossRef]
28. Gaboreau, S.; Robinet, J.C.; Pret, D. Optimization of pore-network characterization of a compacted clay material by TEM and FIB/SEM imaging. *Microporous Mesoporous Mater.* **2016**, *224*, 116–128. [CrossRef]
29. Hemes, S.; Desbois, G.; Urai, J.L.; Schröppel, B.; Schwarz, J.O. Multi-scale characterization of porosity in Boom Clay (HADES-level, Mol, Belgium) using a combination of X-ray μ-CT, 2D BIB-SEM and FIB-SEM tomography. *Microporous Mesoporous Mater.* **2015**, *208*, 1–20. [CrossRef]
30. Houben, L.; Sadan, M.B. Refinement procedure for the image alignment in high-resolution electron tomography. *Ultramicroscopy* **2011**, *111*, 1512–1520. [CrossRef] [PubMed]
31. Ebrahimi, D.; Whittle, A.J.; Pellenq, R.J.M. Effect of polydispersity of clay platelets on the aggregation and mechanical properties of clay at the mesoscale. *Clays Clay Miner.* **2016**, *64*, 335–347. [CrossRef]
32. Ebrahimi, D.; Whittle, A.J.; Pellenq, R.J.M. Mesoscale properties of clay aggregates from potential of mean force representation of interactions between nanoplatelets. *J. Chem. Phys.* **2014**, *140*, 154309. [CrossRef]
33. Ho, T.A.; Greathouse, J.A.; Wang, Y.; Criscenti, L.J. Atomistic structure of mineral nano-aggregates from simulated compaction and dewatering. *Sci. Rep.* **2017**, *7*, 15286. [CrossRef] [PubMed]
34. Coelho, D.; Thovert, J.F.; Adler, P. Geometrical and transport properties of random packings of spheres and aspherical particles. *Phys. Rev. E* **1997**, *55*, 1959–1978. [CrossRef]
35. Jia, T.; Zhang, Y.; Chen, J. Simulation of granular packing of particles with different size distributions. *Comput. Mater. Sci.* **2012**, *51*, 172–180. [CrossRef]
36. Hubert, F.; Bihannic, I.; Prêt, D.; Tertre, E.; Nauleau, B.; Pelletier, M.; Demé, B.; Ferrage, E. Investigating the anisotropic features of particle orientation in synthetic swelling clay porous media. *Clays Clay Miner.* **2013**, *61*, 397–415. [CrossRef]

37. Hassan, M.S.; Villieras, F.; Gaboriaud, F.; Razafitianamaharavo, A. AFM and low-pressure argon adsorption analysis of geometrical properties of phyllosilicates. *J. Colloid Interface Sci.* **2006**, *296*, 614–623. [CrossRef] [PubMed]
38. Chaikin, P.M.; Lubensky, T.C. *Principles of Condensed Matter Physics*; Cambridge University Press: Cambridge, UK, 2000.
39. Eppenga, R.; Frenkel, D. Monte Carlo study of the isotropic and nematic phases of infinitely thin hard platelets. *Mol. Phys.* **1984**, *52*, 1303–1334. [CrossRef]
40. Hermans, P.H.; Platzek, P. Beiträge zur Kenntnis des Deformationsmechanismus und der Feinstruktur der Hydratzellulose. *Kolloid Z.* **1939**, *88*, 73–78. (In German) [CrossRef]
41. Méheust, Y.; Knudsen, K.D.; Fossum, J.O. Inferring orientation distributions in anisotropic powders of nano-layered crystallites from a single two-dimensional WAXS image. *J. Appl. Crystallogr.* **2006**, *39*, 661–670. [CrossRef]
42. Meng, L.; Jiao, Y.; Li, S. Maximally dense random packings of spherocylinders. *Powder Technol.* **2016**, *292*, 176–185. [CrossRef]
43. Perdigon-Aller, A.C.; Aston, M.; Clarke, S.M. Preferred orientation in filtercakes of kaolinite. *J. Colloid Interface Sci.* **2005**, *290*, 155–165. [CrossRef] [PubMed]
44. Callahan, J. A nontoxic heavy liquid and inexpensive filters for separation of mineral grains. *J. Sediment. Res.* **1987**, *57*, 765–766. [CrossRef]
45. Gregory, M.R.; Johnston, K.A. A nontoxic substitute for hazardous heavy liquids—aqueous sodium polytungstate ($3Na_2WO_4{\cdot}9WO_3{\cdot}H_2O$) solution (Note). *N. Z. J. Geol. Geophys.* **1987**, *30*, 317–320. [CrossRef]
46. Kak, A.; Slaney, M. *Principles of Computerized Tomographic Imaging*; SIAM: Philadelphia, PA, USA, 2001.
47. Russ, J.C. *The Image Processing. Handbook*, 6th ed.; CRC Press: Boca Raton, FL, USA, 2011.
48. Soille, P. *Morphological Image Analysis: Principles and Applications*; Springer: Berlin, Germany, 2004.
49. Cleary, P.W.; Sawley, M.L. DEM modelling of industrial granular flows: 3D case studies and the effect of particle shape on hopper discharge. *Appl. Math. Model.* **2002**, *26*, 89–111. [CrossRef]
50. Fraige, F.Y.; Langston, P.A.; Chen, G.Z. Distinct element modelling of cubic particle packing and flow. *Powder Technol.* **2008**, *186*, 224–240. [CrossRef]
51. Li, J.; Langston, P.A.; Webb, C.; Dyakowski, T. Flow of sphero-disc particles in rectangular hoppers—A DEM and experimental comparison in 3D. *Chem. Eng. Sci.* **2004**, *59*, 5917–5929. [CrossRef]
52. Kim, J.H.; Ochoa, J.A.; Whitaker, S. Diffusion in anisotropic porous media. *Transp. Porous Media* **1987**, *2*, 327–356. [CrossRef]
53. Mammar, N.; Rosanne, M.; Prunet-Foch, B.; Thovert, J.F.; Tevissen, E.; Adler, P. Transport properties of compact clays: I. Conductivity and permeability. *J. Colloid Interface Sci.* **2001**, *240*, 498–508. [CrossRef] [PubMed]
54. Cavallaro, G.; Lazzara, G.; Milioto, S.; Palmisano, G.; Parisi, F. Halloysite nanotube with fluorinated lumen: Non-foaming nanocontainer for storage and controlled release of oxygen in aqueous media. *J. Colloid Interface Sci.* **2014**, *417*, 66–71. [CrossRef] [PubMed]
55. Lisuzzo, L.; Cavallaro, G.; Parisi, F.; Milioto, S.; Lazzara, G. Colloidal stability of halloysite clay nanotubes. *Ceram. Int.* **2018**, in press. [CrossRef]
56. Ketcham, R.A.; Slottke, D.T.; Sharp, J.M., Jr. Three-dimensional measurement of fractures in heterogeneous materials using high-resolution X-ray computed tomography. *Geosphere* **2010**, *6*, 499–514. [CrossRef]
57. Bunge, H.J. *Texture Analysis in Materials Science: Mathematical Methods*; Elsevier: Amsterdam, The Netherlands, 2013.
58. Labarthet, F.L.; Buffeteau, T.; Sourisseau, C. Orientation distribution functions in uniaxial systems centered perpendicularly to a constraint direction. *Appl. Spectrosc.* **2000**, *54*, 699–705. [CrossRef]
59. Cousin, I.; Levitz, P.; Bruand, A. Three-dimensional analysis of a loamy-clay soil using pore and solid chord distributions. *Eur. J. Soil Sci.* **1996**, *47*, 439–452. [CrossRef]
60. Levitz, P.; Tchoubar, D. Disordered porous solids: From chord distributions to small angle scattering. *J. Phys. I* **1992**, *2*, 771–790. [CrossRef]
61. Rozenbaum, O. 3-D characterization of weathered building limestones by high resolution synchrotron X-ray microtomography. *Sci. Total Environ.* **2011**, *409*, 1959–1966. [CrossRef] [PubMed]
62. Torquato, S.; Lu, B. Chord-length distribution function for two-phase random media. *Phys. Rev. E* **1993**, *47*, 2950. [CrossRef]

63. Davidson, P.; Petermann, D.; Levelut, A.M. The measurement of the nematic order parameter by X-ray scattering reconsidered. *J. Phys. II* **1995**, *5*, 113–131. [CrossRef]
64. Lemaire, B.; Panine, P.; Gabriel, J.; Davidson, P. The measurement by SAXS of the nematic order parameter of laponite gels. *EPL Europhys. Lett.* **2002**, *59*, 55. [CrossRef]
65. Sanchez-Castillo, A.; Osipov, M.A.; Giesselmann, F. Orientational order parameters in liquid crystals: A comparative study of X-ray diffraction and polarized Raman spectroscopy results. *Phys. Rev. E* **2010**, *81*, 021707. [CrossRef] [PubMed]

Article

Development of a Zeolite A/LDH Composite for Simultaneous Cation and Anion Removal

Breno Gustavo Porfírio Bezerra [1], Lindiane Bieseki [2], Djalma Ribeiro da Silva [1] and Sibele Berenice Castellã Pergher [1,2,*]

1 Posgraduate Program in Chemistry, Chemistry Institut, Federal University of Rio Grande do Norte. Av Senador Salgado Filho, 3000. CEP 59078-970 Natal/RN, Brazil; brenogpb@gmail.com (B.G.P.B.); djalmarib@gmail.com (D.R.d.S.)

2 Molecular Sieves Laboratory, Chemistry Institut. Av Senador Salgado Filho, 3000. CEP 59078-970 Natal/RN, Brazil; lindiane.bieseki@gmail.com

* Correspondence: sibelepergher@gmail.com

Received: 25 January 2019; Accepted: 18 February 2019; Published: 22 February 2019

Abstract: Wastewater from the oil industry is a major problem for aqueous environments due to its complexity and estimated volume of approximately 250 million barrels per day. The combination of these petroleum pollutants creates risks to human health, and their removal from the environment is considered a major problem in the world today. Thus, this work has the objective of studying the treatment of this type of effluent through the adsorption method using the following exchange materials: cationic, anionic, their combination by a sequential method, and a composite material. Zeolite A, a layered double hydroxide (LDH), and the new composite material formed by zeolite A and LDH structures were synthesized for this study. All were used for the simultaneous treatment of cations and anions in a complex sample such as water produced from petroleum production. The composite demonstrated an excellent ability to simultaneously remove cations and anions. The results obtained after the different treatment modes of the effluent using different materials varied from 85% to 100% for the removal of cations and from 56% to 99.7% for the removal of anions.

Keywords: water produced; adsorbent materials; composite

---

1. Introduction

The regular operations of the oil and gas industry are characterized by a large amount of water injected to facilitate the recovery of oil. This water is brought to the surface along with hydrocarbons (oil and gas), salt and other solutes, and it is commonly known as "produced water" [1]. The liquid waste stream produced by the oil industry is estimated to be approximately 250 million barrels per day, and the water to oil ratio is at least 3:1. The complex composition of the produced water (PW) is variable and depends on the natural geological characteristics of the location. Its properties can vary, as it contains several soluble mineral ions and is often of an acidic nature [2]. Some constituents of concern in PW are salts (expressed as salinity), total dissolved solids, oil and grease content, natural inorganic and organic compounds (e.g., chemicals that cause hardness and scaling such as calcium, magnesium, sulphates and barium) and the chemical additives used in drilling, fracturing and well operation, which may have some toxic properties (e.g., biocides and corrosion inhibitors) [3]. The inorganic minerals, which are present as dissolved salts, dissolved in the PW are strongly related to the geochemical characteristics of the well, and they are the cations Na^+, K^+, Ca^{2+}, Mg^{2+}, Ba^{2+}, Sr^{2+}, and Fe^{2+} as well as the anions Cl^-, $SO_4{}^{2-}$, $CO_3{}^{2-}$, and $HCO_3{}^-$, which affect the PW chemistry in terms of the buffering capacity, salinity and potential for scaling [4]. The toxic metals commonly found in PW are cadmium, chromium, copper, lead, mercury, nickel, silver and zinc, and they are mainly of natural origin [4].

Wastewater from the petroleum industry is a major problem for aqueous environments. The combination of these petroleum pollutants creates risks to human health [5]. Their removal from the environment is considered a major problem in today's world [6]. The reuse or recycling of water has become mandatory, especially in countries with water concerns. Many countries have implemented more stringent regulations for the permitted limits of oil and gas (O and G) that can be discharged into the produced water, ranging from 10 mg L^{-1}, according to China's Ministry of Environment, up to the maximum limit of 42 mg L^{-1}, regulated by the United States Environmental Protection Agency (USEPA) [7]. In Brazil, in environmental legislation, the maximum value allowed is up to a 20 mg L^{-1} value defined through CONAMA-National Environment Council, with resolutions no. 357 (CONAMA 357/2005) [8] and no. 430 (CONAMA 430/2011) [9], which establishes the conditions and standards for the discharge of effluents by determining the maximum limits of concentrations of the parameters. New regulations have promoted the development of ecological and economic disposal methods [10]. This can be seen as an opportunity to treat the water produced and provide a viable source of water, which is beneficial in many applications, for which the quality of drinking water is not necessary while avoiding serious environmental damage. Thus, this work has the objective of PW treatment by adsorption method using a composite material that can be capable of removing simultaneously cations and anions.

It is well known that zeolites can remove cations and layered double hydroxides (LDHs) can remove anions. So these materials can be used for PW treatment. In this study, a composite made by zeolite A and LDH was synthesized and used for PW treatment. Previous studies have performed the synthesis of a zeolite A and LDH composite. Yamada et al. [11] coated a sample of zeolite A with LDH through the dripping of a mixture of solutions of magnesium chloride and aluminium chloride but did not study its application. Another study was carried out in order to synthesize a zeolite A and LDH (Mg/Fe) composite using the acid residue of the copper and kaolin processing process for the combined synthesis of zeolite A and LDH. The synthesis occurred in the following two steps: synthesis of LDH by co-precipitation and then the addition of metakaolin for the synthesis of zeolite A [12].

In this study, zeolite A, LDH, and a new composite material formed by zeolite A and LDH (Mg/Al) structures were synthesized separately. The synthesis process of the composite was carried out at room temperature in a shorter synthesis time than in the literature [11,12], with the excellent formation of the crystals. In addition to the new synthesis processes, this study used for the first time the synthesized materials for the simultaneous removal of cations and anions in wastewater produced from petroleum, which is considered a highly complex removal process. Zeolite A is characterized by the presence of molecular-sized pores and cavities, which are occupied, due to charge compensation, by cations and water molecules. These ions are not covalently bound to the structure, which makes the zeolites excellent cation exchangers [13–15]. LDH is an anionic clay with positive layers of di- and trivalent metal and ions and anions in the interlayer region present to neutralize this charge. The freedom of these intercalary anions causes the LDH to have the capacity of an anion exchanger [16,17]. Thus, the composite material is ideal for the reutilization of produced water by the simultaneous removal of the cations and anions present.

2. Materials and Methods

2.1. Synthesis and Characterization of the Materials

For the preparation of the composite based on zeolite A and LDH, two synthetic methodologies were used sequentially. The first methodology was based on the synthesis of IZA for zeolite A [18], and the second methodology was based on the procedure proposed by Climent and co-workers [19], with changes including seeking an LDH with an Mg/Al = 3 ratio. A sample of pure zeolite A and another of LDH were also prepared for comparison with the synthesized composite.

The preparation of the composite consists of two fundamental steps. In the first step, a gel was prepared with the following composition: 0.098 mol SiO_2, 0.049 mol Al_2O_3, 0.157 mol Na_2O and 6.285 mol H_2O. To prepare the gel, 1.266 g of NaOH (Sigma Aldrich) was added to 100 g of Milli-Q H_2O, and after complete dissolution, the solution is divided into two equal volume fractions, called V1 and V2. In the first fraction V1, 9.100 g of sodium aluminate (50%–56% Al_2O_3 40%–45% Na_2O) was added until dissolution. In the second fraction V2, 13.225 g of deionized H_2O plus 6.000 g of NaOH and 5.900 g of SiO_2 (Aerosil 200 Degussa) were added. After the homogenization of both fractions, fraction V1 was rapidly added to fraction V2. After stirring the system for approximately 30 min, the prepared gel was placed in Teflon autoclaves, where it was kept in a hot block for 4 h at 100 °C.

During the crystallization of the gel, a solution of $Mg(NO_3)_2 \cdot 6H_2O$ at 0.225 mol·L^{-1} was prepared with $Al(NO_3)_2 \cdot 9H_2O$ at 0.075 mol·L^{-1} in a final volume of 200 mL (solution A). A 0.4 mol·L^{-1} NaOH solution was also prepared with 0.2 mol L^{-1} Na_2CO_3 to a final volume of 400 mL (solution B).

For the second step, the gel obtained in the first step was transferred to a stirred glass beaker, in which solution (A) containing the magnesium and aluminium nitrates and solution (B) containing the sodium carbonate and sodium hydroxide were dripped in simultaneously. The dripping occurred slowly under agitation on the synthesis gel at a constant temperature of 60 °C. At the end of the addition of solutions A and B, the mixture was stirred for 3 h at 60 °C for ageing. After this time, the solid obtained was separated by filtration and washed with distilled water until pH = 7, and it was finally dried at room temperature. The material produced was denominated ZAHD composite and was also calcined in air with a 3 °C/min heating ramp at 400 °C for 3 h. The composite before and after being calcined was then characterized.

The prepared composite material was characterized by differential thermal and thermogravimetric analysis (DTG, DTA) under a dynamic nitrogen atmosphere with a flow rate of 20 mL min^{-1} at a heating rate of 10 °C·min^{-1} from 30 and 800 °C using a mass of approximately 10 mg deposited in a platinum sample port. The DTG curves were obtained by calculating the first derivative of the TG curves.

The samples synthesized in this work were characterized by a Bruker D2 Phaser diffractometer (Billerica, MA, USA) using Cu radiation (λ = 1.54 Å) with a step size of 0.02°, a current of 10 mA, and a voltage of 30 kV. A Lynxeye detector (192 channels), divergent 0.6 mm slit, 0.1 s time, and 1 mm anti-air scattering screen was used.

In the Chemical analysis, the samples were characterized in a Bruker S2 Ranger apparatus (Billerica, MA, USA), using P power radiation, a Ag power of 50 W, a maximum voltage of 50 kV, a maximum current of 2 mA, and a XFlash®Silicon Drift Detector (Billerica, MA, USA), with the results were normalized to 100%.

The samples were analysed using a ZEISS brand electronic scanning electron microscope (Oberkochen, German), an Auriga model FEG (field emission gun) type emitter, with a voltage of 20 kV, a chemical analysis detector using energy dispersion spectroscopy (EDS, Bruker, Billerica, MA< USA) and a XFlash®410-M detector (Bruker, Billerica, MA< USA).

The specific area analysis and N_2 adsorption isotherm were obtained by the physical adsorption of nitrogen on the material by the Brunanuer-Emmett-Teller method (BET). This method is based on the determination of the volume of N_2 adsorbed at various relative pressures at the temperature of liquid nitrogen, at pressures of up to 2 atm and at relative pressures (p/p_0) of less than 0.3 atm. For the performance of this test, a specific area metre, a Micromeritcs brand ASAP 2020 (Altalnta, GA, USA), was used. The data were taken from the nitrogen sorption isotherms at 77 K to determine the specific surface area (S_{BET} and Gurvich rule for the total pore volume, VTP).

2.2. Treatment of Water Produced from Petroleum

The treatment of the water produced from oil (from the RN CE Operation Unit) was performed using 50 mL of the solution in contact with a mass of 0.30 g of each material. The samples were placed in 150 mL Erlenmeyer flasks and subjected to a stirring orbital shaker table operating at 200 rpm for periods equal to 4 h. At the end of this time, the samples were filtered and analysed by IC and ICP-OES.

The first treatment was with LDH and then with zeolite A, while in the second treatment, the order was reversed with zeolite A being used before LDH. In the last treatment, the composite was used.

Cation and anion concentrations were determined by liquid chromatography (IC) and by inductively coupled plasma optical emission spectrometry (ICP OES). The analyses were performed in triplicate.

The produced water sample was analysed by inductively coupled plasma optical emission spectrometry (ICP OES). The equipment used was an iCAP 6300 Duo model (Thermo Fisher Scientific, Massachusetts, USA) with axial and radial views and a simultaneous CID (charge injection device) Detector. Commercial argon with a purity of 99.996% (White Martins-Praxair) was used along with the following parameters: power source RF at 1150 W, nebulizer gas flow at 0.75 L min^{-1}, auxiliary gas flow at 0.5 L min^{-1}, and stabilization time of 15 s.

The sample of water produced was also analysed by ion chromatography (IC) using an ICS-2000 DIONEX ion chromatograph (Sunnyvale, CA, USA), with an in situ eluent generator, conductivity detector and electrochemical suppression as well as an AS40 DIONEX autosampler. The analytical column and guard column used were an IonPAC AS19 2 × 250 mm and an IonPAC AG19 2 × 50 mm, respectively, both from DIONEX.

3. Results

3.1. Synthesis and Characterization

Figure 1 shows the diffractograms and the structure scheme of the following materials: LDH, zeolite A and zeolite A/LDH composite (ZAHD).

Only the LDH phase is present in the diffractogram of Figure 1, where the characteristic reflections are observed at 2θ = 11.34°, 22.88°, 34.35°, 38.73° and 45.73°, with the value of d(003) = 7.8 Å being due to the presence of carbonate between the lamellae [20,21]. The sample of zeolite A (Figure 1) is pure zeolite A and presents all of the characteristic reflections of this phase, with it then being used as a standard [22]. The diffractogram of the composite (Figure 1) shows characteristic reflections of the LDH and zeolite A compounds, corresponding to the new composite (ZAHD).

Figure 1 shows a scheme of the materials structures and the proposed structure of the composite, that is zeolite A crystals covered by LDH layers (see also SEM results in Figure 3).

A thermogravimetric analysis of the composite was performed with the objective of observing the changes in relation to the loss of mass occurring upon heating. The TG/DTG curves and mass loss percentages for each event are shown in Figure 2.

To confirm the formation of the composite, high resolution scanning electron microscopy (SEM) analyses were performed. LDH, zeolite A and ZAHD composite samples were analysed, and the images are presented in Figure 3.

The micrographs of zeolite A and LDH are in agreement with the literature according to Reference [19]. The zeolite A presents a morphology of cubic crystals with defined edges [23]. For the LDH, a morphology showing a series of hexagonally squamous particles is indicated. In the synthesis of the composite, from the enlargement, it can be observed that it has a well-defined morphology resembling cubic crystals, but its surface has a smooth homogeneous layer, leaving these crystals without a defined format but totally covered.

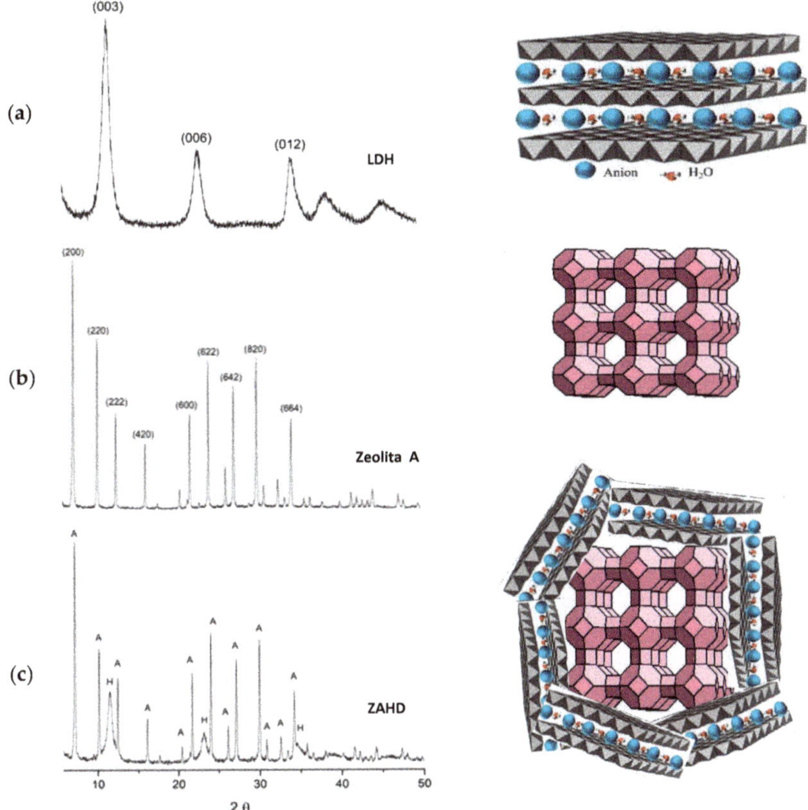

Figure 1. Diffractograms of the structure scheme of the synthesized materials: (**a**) LDH, (**b**) zeolite A, and (**c**) ZAHD composite.

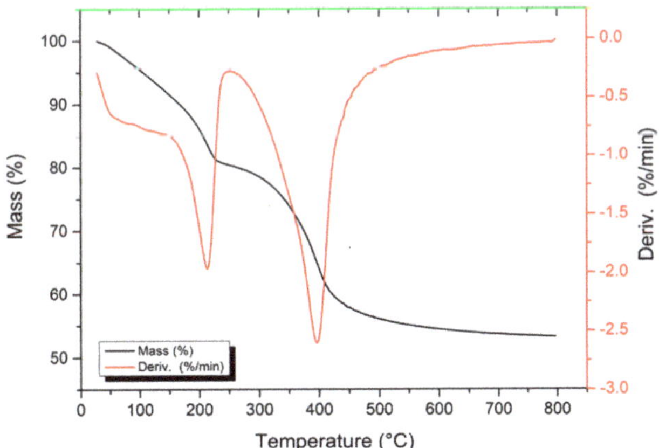

Figure 2. Thermogravimetric and TG/DTG curves of the ZAHD composite.

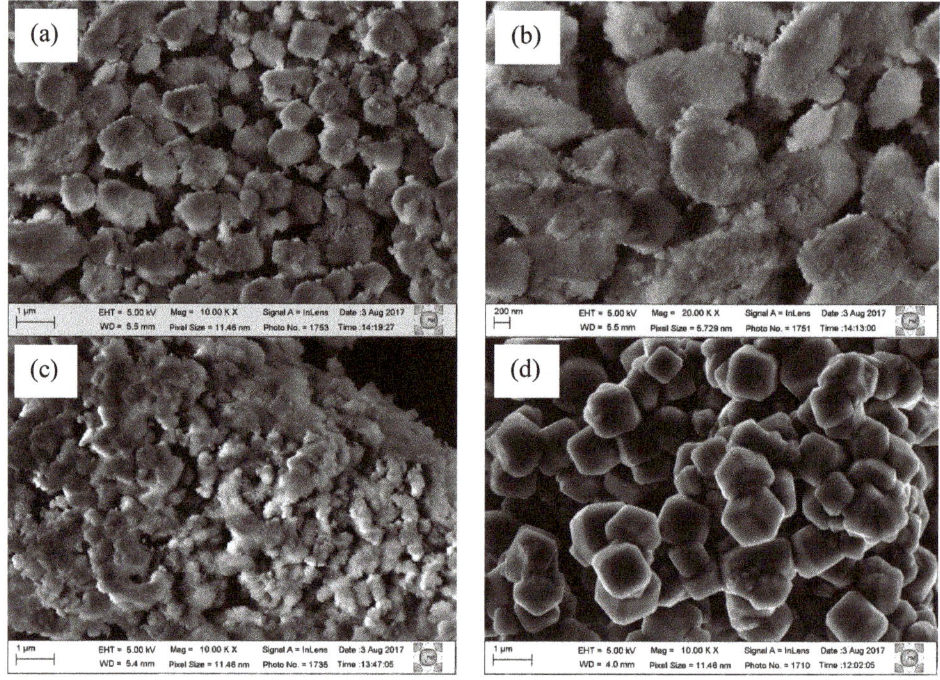

Figure 3. Micrograph of the composite ZAHD (**a**,**b**) material compared to zeolite A (**c**) and LDH (**d**).

Table 1 presents the FRX results for the three studied materials: zeolite A, LDH, and the composite (ZAHD).

Table 1. FRX results in % for the synthesized materials.

Material	Al_2O_3	MgO	SiO_2	Na_2O	Si/Al	Mg/Al
Zeolite A	35.50	1.5	45.35	16.9	1.10	0.05
LDH	37.97	58.21	0	3.06	0	1.9
ZAHD	34.71	20.9	33.96	10.0	0.83	0.77

The chemical composition for zeolite A is in agreement with that observed for this material in the literature [24].

The Mg/Al ratio for the LDH was 1.9 and that value was lower than expected (3.0) The Mg/Al ratio for the composite was 0.77, while the Si/Al ratio was equal to 0.83.

In the determinations of the specific surface areas, the nitrogen adsorption/desorption technique was used at 77 K, the isotherms for the synthesized materials are presented in Figure 4.

Figure 4 shows that the LDH isotherm is of type II, which is characteristic of non-porous materials. This occurs because the lamella dimensions are larger than the relative size of the nitrogen molecule [25,26]. The zeolite A isotherm shown in Figure 4 is classified as being type I, presenting a low specific area due to the size of its micropores [27]. In Figure 4, the composite isotherm is type IV, which is characteristic of porous and lamellar materials. Mathematical models can be applied to determine the specific surface area and total pore volume. Table 2 shows the results obtained.

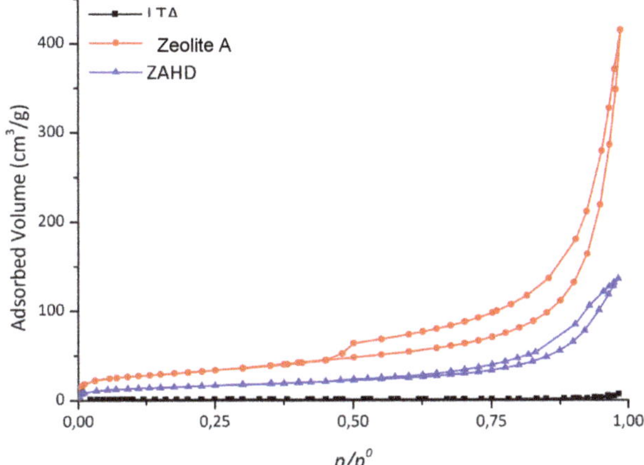

Figure 4. N_2 isotherms of the synthesized samples: LDH, zeolite A, and ZAHD composite.

Table 2. Textural results for the synthesized materials.

Material	S_{BET} (m^2/g)	VTP (cm^3/g)/0.98
Zeolite A	5	0.01
LDH	113	0.60
ZAHD	57	0.21

The BET surface area value and micropore volume calculated for the composite are lower than those obtained for the pure LDH, and this is justified by the fact that a large amount of the composite is formed by zeolite A.

3.2. Treatment of a Sample of Water Produced from Petroleum

Knowing the properties of the LDH, zeolite A and the composite material, in this work the treatment of samples of petroleum produced water with different concentrations of cations and anions was carried out. Table 3 shows these results. The first treatments were using only Zeolite A and LDH; then two treatments performed sequentially form LDH/zeolite A, and the reverse order forms zeolite A/LDH, with the last sample used being the composite. Figure 5 shows the ICP-OES results in relation to the removal of cations for the different treatments employed.

Table 3. Adsorption results for the synthesized materials.

Parameters	Initial Concentration on PW (mg/L)	ZEO (mg/L)	LDH (mg/L)	LDH/ZEO (mg/L)	ZEO/LDH (mg/L)	COMPOSITE (mg/L)
Al	3.913	0	3.930	0.078	0	0
Cd	0.448	0	0.446	0	0	0.059
Cu	0.385	0	0.384	0	0	0
Mn	2.319	0.070	2.322	0.105	0.075	0.740
Ni	0.482	0	0.484	0	0	0.131
Pb	0.437	0	0.435	0	0	0
Zn	2.407	0	0.404	0	0	0.058
Chloride	203.971	200.345	32.965	25.276	28.965	29.138
Bromide	18.656	18.530	3.572	2.982	3.572	8.09
nitrite	13.97	14.02	2.980	2.947	2.98	6.15
nitrate	60.757	60.573	10.239	9.908	7.239	13.731
phosphate	18.304	18.132	2.080	2.019	2.08	0.07
sulphate	19.027	19.123	14.247	10.406	10.647	5.679

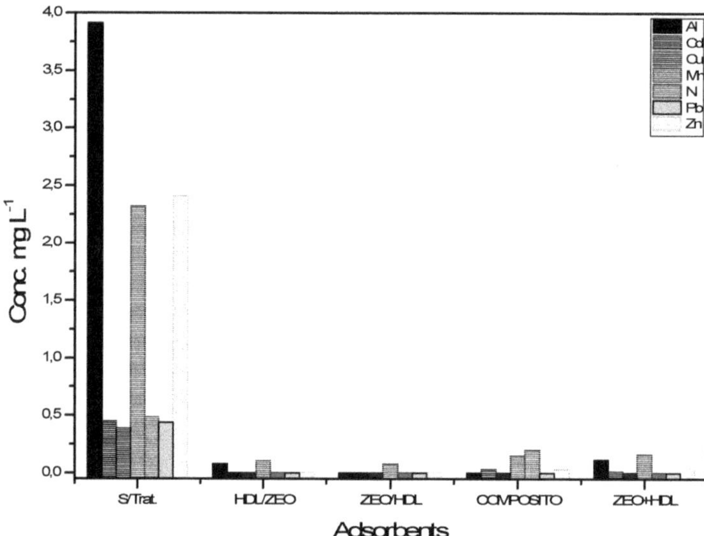

Figure 5. Variations of the concentrations of Al, Cd, Cu, Mn, Ni, Pb and Zn ions in water produced during treatment in the different treatments and analysed by ICP-OES.

The initial concentrations of Al, Cd, Cu, Mn, Ni, Pb and Zn ions without the presence or action of adsorbents were 3.913 mg L^{-1}, 0.448 mg L^{-1}, 0.385 mg L^{-1}, 2.319 mg L^{-1}, 0.482 mg L^{-1}, 0.437 mg L^{-1} and 2.407 mg L^{-1}. The removal percentages obtained in the treatment performed by the composite were 99.2%, 95%, 100%, 85%, 100%, 100% and 98.8% for Al, Cd, Cu, Mn, Ni, Pb and Zn, respectively.

Chloride, bromide, nitrite, nitrate, phosphate and sulphate concentrations present in water produced during treatment with the different treatments, as analysed by IC, are presented in Figure 6.

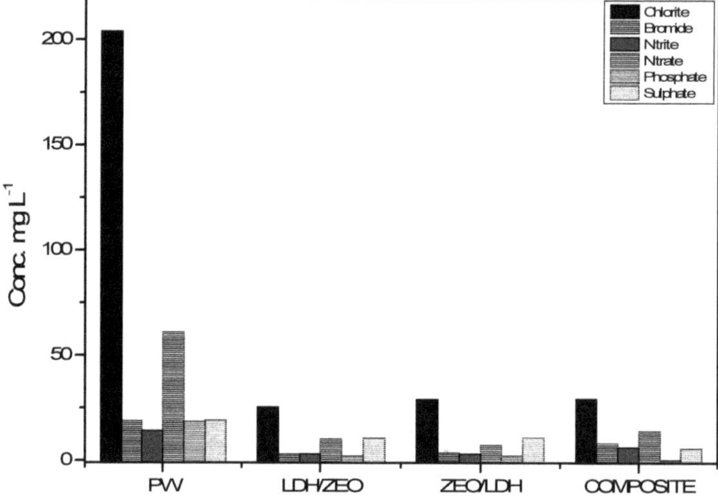

Figure 6. Chloride, bromide, nitrite, nitrate, phosphate and sulphate ion concentrations present in water produced during treatment with the different treatments, as analysed by IC.

The initial concentrations of chloride, bromide, nitrite, nitrate, phosphate and sulphate without the presence or action of adsorbents were 203.971 mg L^{-1}, 18.656 mg L^{-1}, 13.97 mg L^{-1}, 60.757 mg L^{-1},

18.304 mg L^{-1} and 19.027 mg L^{-1}. The removal rates obtained by the composite treatment were 85.7%, 57%, 56%, 77.4%, 99.7% and 70.1% for the chloride, bromide, nitrite, nitrate, phosphate and sulphate ions, respectively.

4. Discussion

Comparing the X-ray diffraction results of the LDH and zeolite A samples with the ZAHD composite sample, it is observed that the reflections present coincide with the two phases mentioned above. It is observed that the intensities of the reflections related to the zeolite A are less intense than the standard material, which may be related to the later stage of formation of the LDH phase, where an alkaline system is used. In the formation of the composite, the crystals of zeolite A are surrounded by the layered double hydroxide. This may also cause the reflections of the zeolite to be less intense. In contrast from the study conducted by Yamada et al. (2006) [11], the reflections of LDH can be clearly identified, especially reflections corresponding to the 003 and 006 planes.

In the thermogravimetric analysis, two mass losses with three events were observed, which can be observed in Figure 2. The first loss of mass occurred between 30–100 °C and was 5.07%, which was attributed to the removal of poorly adsorbed water due to the low temperature. The second mass loss occurring between 100–200 °C represented 15.41% of the total mass. The temperature at which this event occurs suggests that the loss is due to the removal of the interlamellar water due to the presence of LDH and the poorly adsorbed water in the zeolite structure. A third mass loss occurred at approximately 200–400 °C (where the maximum loss occurs), which is within the correct temperature range for dehydroxylation and decarbonation. So this temperature (400 °C) was used for calcination. This is the largest mass loss, accounting for 20.09% of the total mass lost [28]. It is common that the higher mass loss during the thermal decomposition of a present LDH results from the simultaneous dehydroxylation and removal of anhydrous interlayers of [29].

In the micrographs shown in Figure 3, it is observed that the LDH particles are inclined to strongly agglomerate together. The thickness of the squamous particles is very small, only a few nanometres, which has already been observed in other studies [11,12,30]. In our study, due to the composite synthetic process being carried out from the crystallized material but still dispersed in the aqueous synthetic matrix (mother water), a thicker layer was formed around the crystals of zeolite A.

The Si/Al and Mg/Al ratio data presented for the composite in Table 1 are lower than those obtained for zeolite A and LDH, respectively. This value is down because it is estimated using the amount of total aluminium, i.e., the amount present in zeolite A and HDL. The amount of NaO_2 present is lower in the composite compared to pure zeolite A, which may indicate that a portion of the Na present in the zeolite structure may have been replaced by Mg.

The formation of the composite is supported by the adsorption data, where we can observe that the specific surface area obtained was in between those of LDH and zeolite A (Table 2). The value of 113 $m^2 \cdot g^{-1}$ for LDH may be related to the synthetic temperature employed. Lower temperatures favour the formation of LDH-type materials with higher specific surface areas [31].

These materials, Zeolite A, LDH and composite; are materials with charge deficiency. So the mechanism that will occur in the adsorption process will be the exchanging of cations and anions. So, in the case of zeolites, the molar ratio Al/Si is important, because the Al will give a negative charge to the framework, so with more aluminium, more cationic exchange capacity the material will have. And in the case of LDH, the Al gives a positive charge to the framework, so with more Al, more anionic exchange capacity the material will have.

In the produced water (PW) treated only with zeolite A (0.3 g for 4 h), approximately all of the cations are removed (~100%), and the anions were not removed. When the PW was treated only with LDH (3 g for 4 h), approximately 80% to 90% of anions were removed (the only sulphate was ~25%), and the cations were not removed (Table 3). Thus, to simultaneously remove cations and anions from the PW, it will be necessary to employ both materials, LDH and zeolite A. The result showed that the sequential treatments have the same results, independent of the order in which they were used (60 to

100% for anions and 80% to 100% for cations). In these treatments, 0.3 g of each material was used for 4 h, and thus, the total used was 0.6 g for 8 h. When the composite was used, it was 0.6 g for 4 h, and it demonstrated the same capacity for removing cations and anions as the sequential treatment. Thus, the composite material is efficient for removing cations and anions from PW simultaneously and can do so in only 4 h. Another advantage of the composite material is that this material has a morphology and particle sizes more homogeneous than the use of zeolite A and LDH separately, sequentially or mixed. This fact can help to avoid diffusion problems.

5. Conclusions

In this study, the syntheses of zeolite A, layered double hydroxide (LDH), and a composite material based on zeolite A and layered double hydroxide, called composite (ZAHD), were successfully performed at room temperature in only 7 h.

The materials zeolite A, LDH and the composite (ZAHD) were used for the simultaneous treatment of cations and anions in a complex sample of water produced from petroleum, and it was observed from the results that all of the materials have very good adsorption capacities. The results obtained after the different treatment modes of the effluent in different materials varied from 80 to 100% for the removal of cations and from 60% to 100% for the removal of anions. These values are extremely satisfactory considering the complexity of the effluent produced from petroleum, and since the concentrations of cations and anions are in the range of mg L^{-1} to μg L^{-1}, treatment is even more difficult.

Author Contributions: All of the authors worked on this research. Conceptualization, B.G.P.B. and S.B.C.P.; methodology, B.G.P.B. and L.B.; validation, B.G.P.B. and L.B.; formal analysis, B.G.P.B. and D.R.d.S.; writing—original draft preparation, B.G.P.B. and L.B.; writing—review and editing, S.B.C.P. and D.R.d.S.; supervision, S.B.C.P. and D.R.S; project administration, S.B.C.P.

Funding: This research received no external funding.

Acknowledgments: To the RN CE Operation Unit for providing the produced water.

Conflicts of Interest: The authors declare no conflict of interest.

References

1. Xu, J.; Srivatsa Bettahalli, N.M.; Chisca, S.; Khalid, M.K.; Ghaffour, N.; Vilagines, R.; Nunes, S.P. Polyoxadiazole hollow fibers for produced water treatment by direct contact membrane distillation. *Desalination* **2018**, *432*, 32–39. [CrossRef]
2. Khalib, Z.; Verbeek, P. (Shell) Water to Value—Produced Water Management for Sustainable Field Development of Mature and Green Fields. In Proceedings of the SPE International Conference on Health, Safety and Environment in Oil and Gas Exploration and Production, Kuala Lumpur, Malaysia, 20–22 March 2002; p. 4.
3. Arthur, D.J.; Hochheiser, W.H.; Bottrell, M.D.; Brown, A.; Candler, J.; Cole, L.; DeLao, D.; Dillon, L.W.; Drazan, D.J.; Dusseault, M.B.; et al. *Management of Produced Water from Oil and Gas Wells*; NCP North American Resource Development: Washington, DC, USA, 2011.
4. BR, H.; SR, D. Review of Potential Technologies for the Removal of Dissolved Components from Produced Water. *Chem. Eng. Res. Des.* **1994**, *72*, 176–188.
5. Farhadian, M.; Duchez, D.; Vachelard, C.; Larroche, C. Monoaromatics removal from polluted water through bioreactors—A review. *Water Res.* **2008**, *42*, 1325–1341. [CrossRef] [PubMed]
6. Das, N.; Chandran, P. *Microbial Degradation of Petroleum Hydrocarbon Contaminants: An Overview*; Hindaw: NewYork, NY, USA, 2011; Volume 2011, pp. 1–13.
7. *Guidelines Establishing Test Procedures for the Analysis of Oil and Grease and Non-Polar Material under the Clean Water Act and Resource Conservation and Recovery Act*; Final Rule; Environmental Protection Agen: Washington, DC, USA, 1999; Volume 64, pp. 26315–26327.
8. *Resolução CONAMA n° 357, de 17 de março de 2005*; Ministério do Meio Ambiente: Brasília, Brasil, 2005; 23p.
9. *Resolução CONAMA n° 430, de 13 de maio de 2011*; Ministério do Meio Ambiente: Brasília, Brasil, 2011; pp. 1–8.
10. Tellez, G.T.; Nirmalakhandan, N.; Gardea-Torresdey, J.L. Performance evaluation of an activated sludge system for removing petroleum hydrocarbons from oilfield produced water. *Adv. Environ. Res.* **2002**, *6*, 455–470. [CrossRef]

11. Yamada, H.; Watanabe, Y.; Hashimoto, T.; Tamura, K.; Ikoma, T.; Yokoyama, S.; Tanaka, J.; Moriyoshi, Y. Synthesis and characterization of Linde A zeolite coated with a layered double hydroxide. *J. Eur. Ceram. Soc.* **2006**, *26*, 463–467. [CrossRef]
12. Da Silva, L.N.; dos Santos Moraes, D.; Santos, S.C.A.; Corrêa, J.A.M. Joint synthesis of Zeolite A-LDH from mineral industry waste. *Appl. Clay Sci.* **2018**, *161*, 163–168. [CrossRef]
13. Braga, A.A.C.; Morgon, N.H. Descrições estruturais cristalinas de zeólitos. *Quim. Nova* **2007**, *30*, 178–188. [CrossRef]
14. Guisnet, M.; Ribeiro, F.R. *Zeólitos: Um nanomundo ao serviço da catálise*; Fundação Calouste Gulbenkian: Lisbon, Portugal, 2004; ISBN 9789723110715.
15. Melo, C.R.; Riella, H.G. Síntese de zeólita tipo NaA a partir de caulim para obtenção de zeólita 5A através de troca iônica. *Cerâmica* **2010**, *56*, 340–346. [CrossRef]
16. Cavani, F.; Trifirò, F.; Vaccari, A. Hydrotalcite-type anionic clays: Preparation, properties and applications. *Catal. Today* **1991**, *11*, 173–301. [CrossRef]
17. Gomes, J.F.P.; Puna, J.F.B.; Gonçalves, L.M.; Bordado, J.C.M. Study on the use of MgAl hydrotalcites as solid heterogeneous catalysts for biodiesel production. *Energy* **2011**, *36*, 6770–6778. [CrossRef]
18. *Verified Syntheses os Zeolitic Materials*; Mintova, S.; Barrier, N. (Eds.) Synthesis Commission of the International Zeolite Association, 2016; ISBN 978-0-692-68539-6.
19. Climent, M.J.; Corma, A.; Iborra, S.; Epping, K.; Velty, A. Increasing the basicity and catalytic activity of hydrotalcites by different synthesis procedures. *J. Catal.* **2004**, *225*, 316–326. [CrossRef]
20. Fraccarollo, A.; Cossi, M.; Marchese, L. DFT simulation of Mg/Al hydrotalcite with different intercalated anions: Periodic structure and solvating effects on the iodide/triiodide redox couple. *Chem. Phys. Lett.* **2010**, *494*, 274–278. [CrossRef]
21. Radha, A.V.; Kamath, P.V.; Shivakumara, C. Mechanism of the anion exchange reactions of the layered double hydroxides (LDHs) of Ca and Mg with Al. *Solid State Sci.* **2005**, *7*, 1180–1187. [CrossRef]
22. *Collection of Simulated XRD Powder Patterns for Zeolites*; Treacy, M.M.J.; Higgins, J.B. (Eds.) Elsevier: Amsterdam, The Netherlands, 2001; ISBN 978-0-444-53067-7.
23. Maia, A.Á.B.; Neves, R.F.; Angélica, Rô.S.; Pöllmann, H. Synthesis, optimisation and characterisation of the zeolite NaA using kaolin waste from the Amazon Region. Production of Zeolites KA, MgA and CaA. *Appl. Clay Sci.* **2015**, *108*, 55–60. [CrossRef]
24. Rocha Junior, C.A.F.; Santos, S.C.A.; Souza, C.A.G.; Angélica, R.S.; Neves, R.F. Síntese de zeólitas a partir de cinza volante de caldeiras: Caracterização física, química e mineralógica. *Cerâmica* **2012**, *58*, 43–52. [CrossRef]
25. Seftel, E.M.; Niarchos, M.; Mitropoulos, C.; Mertens, M.; Vansant, E.F.; Cool, P. Photocatalytic removal of phenol and methylene-blue in aqueous media using TiO2@LDH clay nanocomposites. *Catal. Today* **2015**, *252*, 120–127. [CrossRef]
26. Pesic, L.; Salipurovic, S.; Markovic, V.; Vucelic, D.; Kagunya, W.; Jones, W. Thermal characteristics of a synthetic hydrotalcite-like material. *J. Mater. Chem.* **1992**, *2*, 1069. [CrossRef]
27. Nascimento, C.R. Síntese da zeólita A utilizando diatomita como fonte de sílicio e alumínio. *Cerâmica* **2014**, *60*, 63–68. [CrossRef]
28. Forano, C.; Hibino, T.; Leroux, F.; Taviot-Guého, C. Layered Double Hydroxides. In *Handbook of Clay Science*; Bergaya, F., Theng, B.K.G., Lagaly, G., Eds.; Elsevier Ltd.: Amsterdam, The Netherlands, 2006; Volume 1, pp. 1021–1095, ISBN 1572-4352.
29. Theiss, F.L.; Palmer, S.J.; Ayoko, G.A.; Frost, R.L. Sulfate intercalated layered double hydroxides prepared by the reformation effect. *J. Therm. Anal. Calorim.* **2012**, *107*, 1123–1128. [CrossRef]
30. Xu, J.; Song, Y.; Zhao, Y.; Jiang, L.; Mei, Y.; Chen, P. Chloride removal and corrosion inhibitions of nitrate, nitrite-intercalated Mg–Al layered double hydroxides on steel in saturated calcium hydroxide solution. *Appl. Clay Sci.* **2018**, *163*, 129–136. [CrossRef]
31. Sharma, S.K.; Kushwaha, P.K.; Srivastava, V.K.; Bhatt, S.D.; Jasra, R.V. Effect of hydrothermal conditions on structural and textural properties of synthetic hydrotalcites of varying Mg/Al ratio. *Ind. Eng. Chem. Res.* **2007**, *46*, 4856–4865. [CrossRef]

Article

Study of Catalytic Combustion of Chlorobenzene and Temperature Programmed Reactions over CrCeOx/AlFe Pillared Clay Catalysts

Yingnan Qiu [1], Na Ye [1], Danna Situ [1], Shufeng Zuo [1,*] and Xianqin Wang [2,*]

1 Zhejiang Key Laboratory of Alternative Technologies for Fine Chemicals Process, Shaoxing University, Shaoxing 312000, China; jadeqyn@163.com (Y.Q.); yena@usx.edu.cn (N.Y.); situdn@163.com (D.S.)

2 Department of Chemical, Biological and Pharmaceutical Engineering, New Jersey Institute of Technology, Newark, NJ 07102, USA

* Correspondence: sfzuo@usx.edu.cn (S.Z.); xianqin.wang@njit.edu (X.W.)

Received: 15 January 2019; Accepted: 25 February 2019; Published: 2 March 2019

Abstract: In this study, both AlFe composite pillaring agents and AlFe pillared clays (AlFe-PILC) were synthesized via a facile process developed by our group, after which mixed Cr and Ce precursors were impregnated on AlFe-PILC. Catalytic combustion of organic pollutant chlorobenzene (CB) on CrCe/AlFe-PILC catalysts were systematically studied. AlFe-PILC displayed very high thermal stability and large BET surface area (S_{BET}). After 4 h of calcination at 550 °C, the basal spacing (d_{001}) and S_{BET} of AlFe-PILC was still maintained at 1.91 nm and 318 m^2/g, respectively. Large S_{BET} and d_{001}-value, along with the strong interaction between the carrier and active components, improved the adsorption/desorption of CB and O_2. When the desorption temperatures of CB and O_2 got closer to the CB combustion temperature, the CB conversion could be increased to a higher level. CB combustion on CrCe/AlFe-PILC catalyst was determined using a Langmuir–Hinshelwood mechanism. Adsorption/desorption/oxidation properties were critical to design highly efficient catalysts for CB degradation. Besides, CrCe/AlFe-PILC also displayed good durability for CB combustion, whether in a humid environment or in the presence of volatile organic compound (VOC), making the catalyst an excellent material for eliminating chlorinated VOCs.

Keywords: AlFe-pillared clay; CrCeOx; chlorobenzene; catalytic combustion; temperature-programmed reaction

1. Introduction

Chlorinated volatile organic compounds (CVOCs) are considered to be very harmful to the environment, not only a direct harm on human health but also destroy the ozone layer [1,2]. Today, the major industrial processes for CVOCs elimination involve direct combustion at very high temperatures (above 850 °C). This is a fairly expensive process and produces highly toxic byproducts or intermediates by incomplete combustion, such as dioxins, Cl_2, and CO [2,3]. The low operating temperatures (<500 °C) and high selectivity into harmless product, make catalytic combustion an attractive option [4–6].

Due to the high toxicity of dioxins and the need for laboratory safety, model reagents, such as chlorobenzene (CB), are used to predict destruction behavior of dioxins on different catalysts [7,8]. Vanadia-based catalysts [9,10], precious metals (Pt, Pd, Ru) supported on zeolites [11,12], and various oxides [13,14] are employed for the catalytic combustion process. However, these catalysts often have some disadvantages of relative low catalytic performance, rapid deactivation caused by coking or chlorine poisoning, high price, and the formation of polychlorinated benzene [14]. Transition metal oxide catalysts including cobalt, copper, manganese, and chromium oxides can not only resist deactivation caused by chlorine poisoning but also enhance catalytic activities by the modification with rare earth elements [15,16]. Doping ceria into transition metals oxides could improve redox property of metal oxides, increase the mobility of oxygen and the rate of chlorine removal or transfer, thereby enhancing their catalytic properties and reaction stability during CVOCs combustion [17]. The catalytic behavior is associated with the interaction between the catalyst and the reactant, thus, understanding the relationship between adsorption/desorption and combustion of CVOCs over the catalysts becomes very important. The information can not only explain their own catalytic characteristics, but also provide insights into catalytic macroscopic behaviors [18–20].

Catalytic combustion is a typical gas–solid, two-phase reaction, which mostly occurs on the surface of catalysts. The advantage of heterogeneous catalysis is that the catalyst does not need to be separated, so the process can be operated in uninterrupted flow. Nevertheless, if there is a need to separate the catalyst, it can be done in a much simpler way than a homogeneous catalyst. Heterogeneous catalysts have been well applied in many fields such as heterogeneous Suzuki cross-coupling reaction catalyzed by magnetically recyclable nanocatalyst [21], modulated large-pore mesoporous silica as an efficient base catalyst for a Henry reaction [22], and magnetically separable and sustainable nanostructured catalysts for heterogeneous reduction of nitroaromatics [23–25]. The disadvantage of heterogeneous catalysis is that the catalyst can only use the catalytic active points on its surface, which is slightly inefficient, but can be improved by increasing the specific surface area (S_{BET}). Therefore, the active components are usually dispersed on the carrier with large S_{BET}. In a catalytic combustion reaction, the carrier can not only support the dispersed active components, but also increase the stability, selectivity, and activity of catalysts. The support can also reduce the use of high-priced active components, thereby reducing the cost of the catalyst. Common metal oxide catalyst carriers include γ-Al_2O_3, TiO_2, SiO_2, ZrO_2, or their complexes. Molecular sieve supported catalysts also show good catalytic combustion activity. The molecular sieve carriers studied include ZSM-5, β-molecular sieve, SBA-15, MCM-41, and so on.

Natural clay is a hydrated aluminosilicate, which can be designed by crosslinking or pillaring, which leads to the formation of a class of material known as pillared clays (PILCs). Moreover, because of its wide distribution, abundant reserves, and low price, it has its own advantages in replacing existing catalyst carriers. PILCs have large S_{BET}, uniform pore size distribution, and high thermal stability. They are good catalytic materials and carriers. A popular research topic in recent years is synthesizing PILCs supported catalysts for VOCs catalytic combustion reactions [26–28], whereas the application for CVOCs combustion is rare and the structure-activity relationship was not clear. It has been reported in the literature that S_{BET}, pore volume (V_p), and thermal stability of PILC can be improved by the synthesis composite pillaring agents, and these mixed pillaring agents has been widely used in the past 30 years [29]. At present, many kinds of mixed pillaring agents have been synthesized, with aluminum pillaring agents being one of the components. However, defects in the preparation process still exist. Therefore, there is an urgent need to optimize involved unit operations and simplify procedures, especially to reduce the amounts of NaOH and $AlCl_3$ solutions.

The present work intends to simplify the synthesis steps, using a high-temperature hydrothermal one-step method, to prepare AlFe composite pillaring agents, and then synthesize AlFe-PILC [30]. Compared with Na-Mt, AlFe-PILC had good structural characteristics, such as larger S_{BET} and V_p, so it had the potential for use as a catalytic support. How to improve the efficiency of low temperature catalytic combustion of CB and improve the stability of the catalysts are the key problems to be solved in the current catalytic combustion process. This paper intends to use AlFe-PILC as a carrier to prepare CrCeOx catalysts, and to study their application in the catalytic combustion reaction of CB, aiming at forming achievements in advanced catalytic materials and their applications. The adsorption/desorption properties and catalytic performance of CB were systematically studied in order to get a clear map about the structure–activity relationship for the catalytic reaction and explore its potential for further industrial application.

2. Experimental Section

2.1. AlFe-PILC and Catalyst Preparation

The starting material was the sodium form of montmorillonite (Na-Mt) (>100 mesh, Hengsheng Trading Co., Ltd., Baotou, China). AlFe pillaring agents were prepared with Locron L (Clariant, Switzerland, $Al_2(OH)_5Cl{\cdot}2\text{-}3H_2O$) and ferric nitrate solution. The following preparation of AlFe pillaring (the molar ratio of Al/Fe is 5) and AlFe-PILC using the similar method detailed in our previous research [30].

Cr/Na-Mt, CrCe(5:1)/Na-Mt, Ce/AlFe-PILC, Cr/AlFe-PILC, and CrCe/AlFe-PILC were synthesized by impregnating a Cr and Ce nitrate solution onto the equal volumes of supports (Na-Mt and AlFe-PILC) overnight, followed sequentially by drying and calcination at 500 °C for 2 h. Cr, Ce, or CrCe loading for each catalyst was 10 wt.%, and Cr/Ce molar ratios were adjusted to 2.5:1, 5:1, 7.5:1, and 10:1, respectively. All the reagents were analytically pure, and obtained by Shanghai Chemical Reagent Factory (Shanghai, China).

2.2. Characterization

The samples were characterized using X-ray diffraction (XRD) (PANalytical, Almelo, Netherlands) for d_{001} value and phase composition. The S_{BET}, mesopore area (A_{mes}), V_{p}, micropore volume (V_{mic}), and pore size distribution of the samples were determined via N_2 adsorption isotherms using a TristarII 3020 apparatus (Micromeritics Company, Atlanta, GA, USA). High-resolution transmission electron microscopy (HRTEM) on a JEM-2100F (JEOL, Valley, Japan) was employed to get the catalyst morphology and particle size. The chemical compositions of the catalysts were determined with energy dispersive X-ray spectroscopy (EDS) using an Oxford INCA instrument (Oxford Instruments, Warrington, UK). All the characterization methods for the samples have been reported and detailed in our previous research [30–32].

2.3. Catalytic Performance Tests and Temperature-Programmed Reactions

The activity of the catalysts was evaluated in a WFS-3010 microreactor (Xianquan, Tianjin, China). The degradation products were detected by mass spectrometry (MS, QGA, Hiden, U.K.). No byproduct other than H_2O, CO_2, and HCl was detected. Thus, the conversion was calculated based on CB consumption [31]. To further study the "mixture effect" of the feed gases, 1% (v/v) water vapor and 100 ppm toluene were also introduced. Besides, the durability of CrCe (5:1)/AlFe-PILC for the catalytic combustion of CB was investigated at a CB concentration of 500 ppm and gas hourly space velocity (GHSV) of 25,000 h^{-1}.

H_2 temperature-programmed reduction (H_2-TPR) was conducted on a CHEMBET-3000 instrument (Quantachrome, Boynton Beach, FL, USA) to evaluate the reducibility of the catalysts. The sample (50 mg) was pre-treated in air at 300 °C for 0.5 h, and then the temperature was reduced to 100 °C. The flow rate of the reductive gas (5 vol.% H_2/Ar, purified by deoxidizer and silica gel) was 40 mL/min, and the reaction temperature was elevated by 7.5 °C/min. The H_2 uptake was determined using a thermal conductivity detector (TCD) detector (Shimadzu, GC-14C, Kyoto, Japan), and the H_2O produced was absorbed using 5 Å zeolite [32].

Temperature-programmed desorption (TPD) and temperature-programmed surface reaction (TPSR) measurement were carried out in the same equipment as the catalytic performance tests to determine the adsorption capacity and the relationship between desorption performance and catalytic combustion properties [30,32]. Prior to the measurement, 350 mg catalyst was pre-treated in Ar (99.99%) at 300 °C for 30 min, then the temperature was decreased to 50 °C. Adsorption gas (40 mL/min) was a mixture of Ar (99.99%) and CB (about 500 ppm). The quantitative amounts were estimated by integrating the desorption curve. After the adsorption reached an equilibrium (CB concentration in the effluent gas was monitored using Gas Chromatography-Mass Spectrometry (GC-MS), the catalysts were purged by Ar (99.99%) for a period of time at 50 °C until CB concentration to constant. Then, the desorption and catalytic properties of the catalysts were measured from 50 to 500 °C with a heating rate of 7.5 °C/ min in 20 vol.%O_2/80 vol.%Ar (without CB). The reactants and products (such as CB (m/z =112), CO_2 (44), H_2O (18), Cl_2 (71), and HCl (36.5) were analyzed with an on-line MS apparatus.

O_2 temperature-programmed desorption (O_2-TPD) was also performed using the same apparatus. The catalyst (350 mg) was firstly treated in 10 vol.%O_2/90 vol.%Ar at 300 °C for 0.5 h. After the temperature was slowly cooled down to room temperature, followed by an Ar purge (40 mL/min) for 30 min, the sample was heated from 50 to 900 °C with a heating rate of 7.5 °C/min in Ar flow. The signal of desorbed oxygen was monitored by the MS.

3. Results and Discussion

3.1. Material Textural Properties

3.1.1. XRD Analysis

Figure 1 presented the XRD patterns (2θ: 10–80°) of the Cr/Ce catalysts supported by Na-Mt and AlFe-PILC. The diffraction peaks belonging to cristobalite and quartz appear at 19.8° and 26.7° (2θ), respectively [33]. Cristobalite and quartz are two of the main components of montmorillonite. They have the characteristics of high temperature resistance, which can ensure the stability of the catalysts in a high temperature gas–solid continuous reaction. Therefore, they play an important role in catalyst components. The diffraction peaks of Fe_2O_3 appeared in all the AlFe-PILC based catalysts because the amount of Fe_2O_3 increased after AlFe pillaring process. The diffraction peaks of CeO_2 appeared in Ce/AlFe-PILC catalyst. Compared with Cr/Na-Mt, the diffraction peak intensity of Cr_2O_3 in Cr/AlFe-PILC clearly decreased, and the result showed that the dispersion of Cr_2O_3 particles on AlFe-PILC was greatly improved. After adding Ce, the diffraction intensity of Cr_2O_3 for CrCe(5:1)/AlFe-PILC further decreased. On the one hand, the addition of Ce was beneficial to the dispersion of Cr_2O_3, and on the other hand, it may have been due to the reduction of Cr_2O_3 content. Notably, the CeO_2 diffraction peaks were not found in CrCe(5:1)/AlFe-PILC, possibly because the small amount of CeO_2 was highly dispersed on AlFe-PILC.

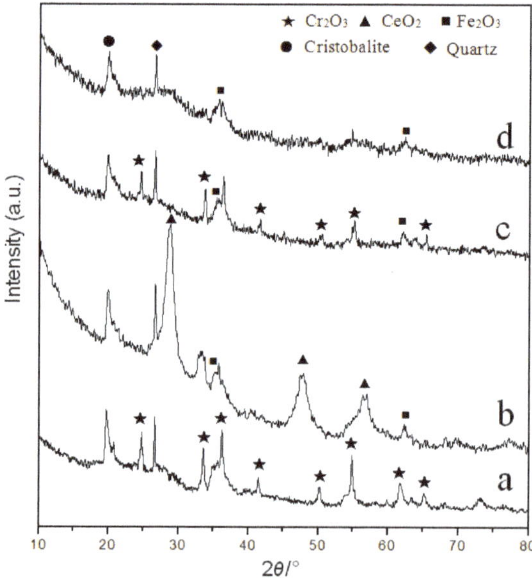

Figure 1. XRD patterns of the catalysts: (**a**) Cr/Na-Mt, (**b**) Ce/AlFe-PILC, (**c**) Cr/AlFe-PILC, and (**d**) CrCe(5:1)/AlFe-PILC.

3.1.2. N_2 Adsorption/Desorption

Table 1 summarizes the textural properties of samples. S_{BET} and A_{mes} of Na-Mt were only 51 and 41 m^2/g, and the values of V_p and V_{mic} were 0.076 and 0.0043 cm^3/g, respectively. AlFe-PILC's S_{BET} and V_p reached 318 m^2/g and 0.195 cm^3/g, respectively, indicating that the formed composite AlFe polycation was relatively large and thus the clay layers were further stripped to form more porous structures. The V_{mic} of AlFe-PILC was 0.077 cm^3/g, and it was about 39.5% in V_p. Compared with the Na-Mt and AlFe-PILC support, the supported Cr or CrCe catalysts exhibited lower S_{BET} and V_p values, indicating that some of the Cr and Ce ions migrated into the pores and clay layers, and thus blocked some of the pores. It was worth noting that a large number of micro-mesoporosity in AlFe-PILC support was favorable for good dispersion of active species and rapid diffusion of reactants, thus significantly enhanced their catalytic activity of various reactions.

Table 1. Characteristics of the samples: values of surface area and pore volume.

Samples	S_{BET} (m^2/g)	A_{mes} [a] (m^2/g)	V_p [b] (cm^3/g)	V_{mic} [c] (cm^3/g)
Na-Mt	51	41	0.076	0.0043
CrCe(5:1)/Na-Mt	22	22	0.058	-
AlFe-PILC	318	168	0.195	0.077
Cr/AlFe-PILC	221	71	0.165	0.066
CrCe(5:1)/AlFe-PILC	13	82	0.156	0.026

[a] Calculated from BJH method. [b] Total pore volume estimated at P/P_0 (relative pressure) = 0.99. [c] Calculated from the *t*-plot method.

In Figure 2a, N_2 adsorption/desorption isotherms for all the materials were type IV, while its type H3 adsorption–desorption hysteresis appeared at P/P_0 above 0.45, indicating that the material had a mesoporous structure and the pores in the material were slit pores formed by layer-like structures. The adsorption amount of Na-Mt was low; however, AlFe-PILC had a pronounced increase in adsorption because more pores were formed by AlFe polyoxycations. The addition of Cr_2O_3 and CeO_2 to Na-Mt and AlFe-PILC decreased N_2 adsorption capacity and thus pore volume, indicating the doped cations entered and/or blocked the pores of Na-Mt and AlFe-PILC. In Figure 2b, the average mesoporous diameters of AlFe-PILC materials were distributed in a narrow range of 3.96 nm and were wider than the pore-diameter distribution range of Na-Mt (3.10 nm), confirming the pore size was increased after pillaring. The average pore size of CrCe(5:1)/AlFe-PILC was in a narrow region of approximately 3.65 nm. The stability of AlFe-PILC support was good and the mesoporous structure was not destroyed.

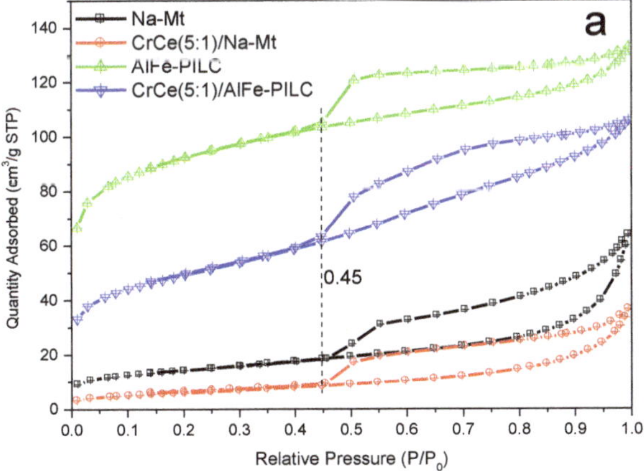

Figure 2. *Cont.*

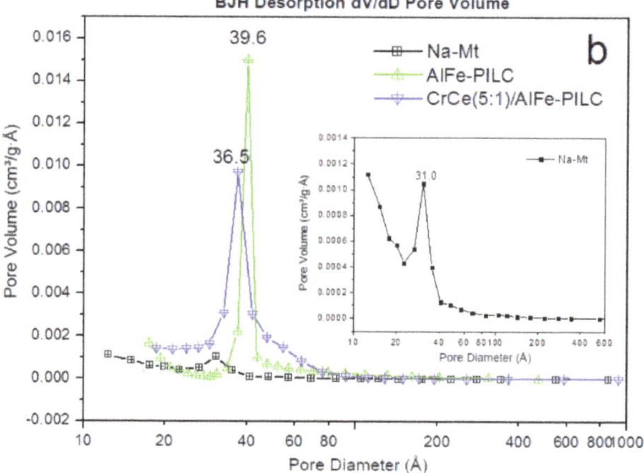

Figure 2. Characteristics of the samples: (**a**) N_2 adsorption/desorption isotherms, and (**b**) pore-size distributions.

3.1.3. HRTEM Analysis

Figure 3 shows HRTEM picture and the EDS spectra of CrCe(5:1)/AlFe-PILC. It can be seen that CrCe(5:1)/AlFe-PILC had a layered structure and the active particles (5–10 nm in size) were uniformly distributed throughout the support, and the layered structure of AlFe-PILC was not damaged after loading active ingredients. Al, Fe, Cr, Ce, O, and other elements were identified in the EDS spectra, which confirmed that the active species (Cr and Ce) were successfully loaded on the surface of AlFe-PILC. The results indicated that AlFe-PILC was a good support for highly dispersed active species. All these properties were conducive to improving the catalytic degradation of CB.

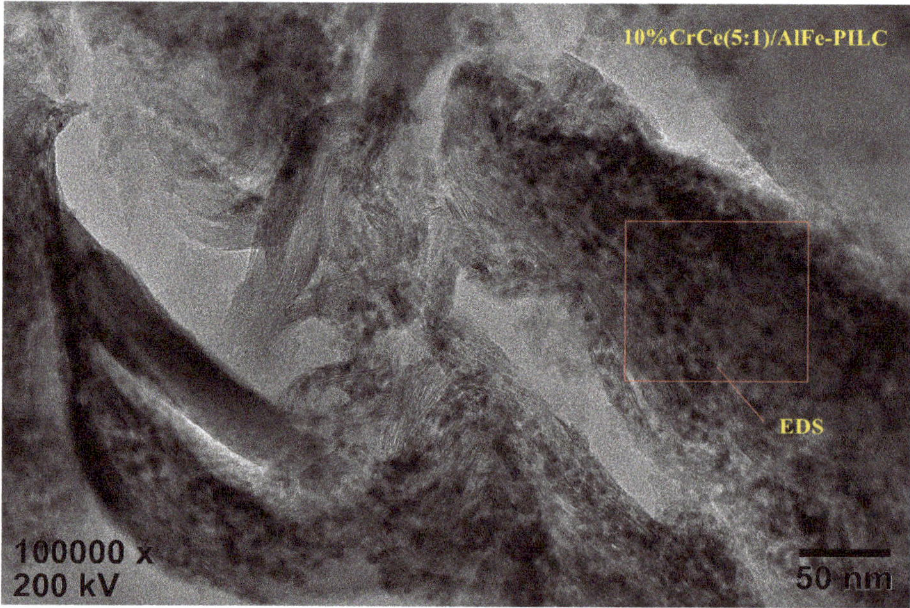

Figure 3. *Cont.*

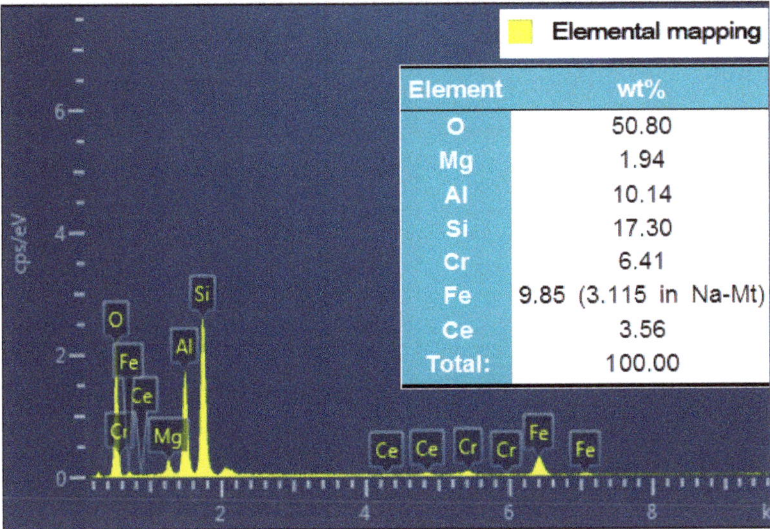

Figure 3. HRTEM picture and the EDS spectra of CrCe(5:1)/AlFe-PILC.

3.2. Catalytic Performance Test

3.2.1. CB Combustion and Durability Test

Figure 4 presents the conversions of CB combustion on various catalysts. In Figure 4a, Cr/Na-Mt exhibited poor performance and did not fully convert CB until 460 °C. Cr/AlFe-PILC caused complete degradation of CB at 320 °C, about 140 °C lower than the degradation temperature of CB required for Cr/Na-Mt. The conversion of Ce/AlFe-PILC was negligibly low, and it was 87%, even at a reaction temperature of 500 °C. Ceria doping significantly improved the catalytic activities of Cr/Na-Mt and Cr/AlFe-PILC. In addition, the molar ratios of Cr/Ce (2.5, 5, 7.5, and 10) had an effect on the catalytic performance of CrCe/AlFe-PILC (Figure 4b). The catalysts exhibited a lower performance when the Cr/Ce ratio was less than 5, possibly indicating Cr_2O_3 was the active species and CeO_2 acted as an assistant. The catalyst performance decreased when the Cr/Ce molar ratio was larger than 5, possibly because less oxygen vacancies existed with a relatively lower amount of CeO_2. Therefore, the content of CeO_2 was one of the key factors to improving the performance of CrCe/AlFe-PILC. In particular, CrCe(5:1)/AlFe-PILC had the highest catalytic performance and could completely degrade CB at about 290 °C. No Cl_2 or other byproducts were detected, showing that the catalyst had good selectivity for HCl without producing secondary pollution.

Figure 5 shows the curves of CB over CrCe(5:1)/AlFe-PILC in the continuous reaction process. There was no significant drop for catalytic activities within 1000 h tests, suggesting that the CrCe(5:1)/AlFe-PILC catalyst was durable. Moreover, this catalyst also displayed good catalytic performances in the presence of 100 ppm toluene or 1% water vapor, further indicating its high potential for industrial application.

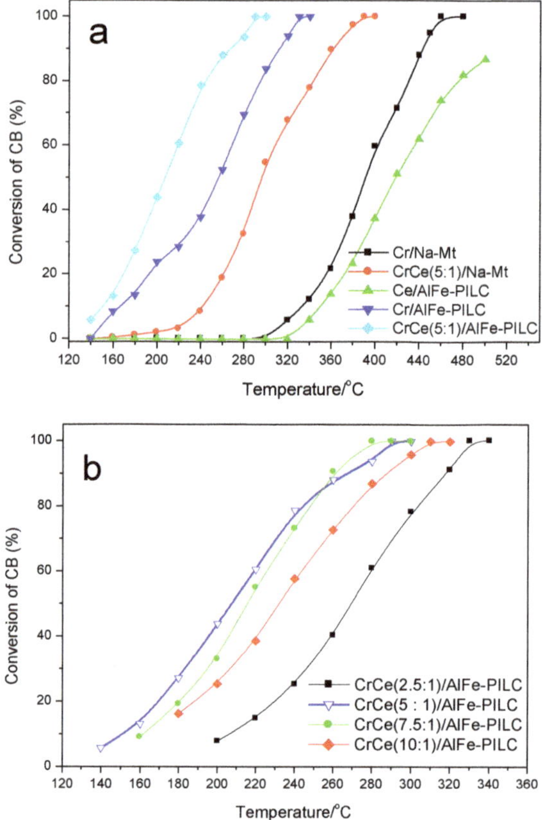

Figure 4. (**a**) CB conversions vs. temperature over Cr/Na-Mt, CrCe(5:1)/Na-Mt, Ce/AlFe-PILC, Cr/AlFe-PILC, and CrCe(5:1)/AlFe-PILC. (**b**) The effect of Cr/Ce molar ratios on CB catalytic combustion over CrCe/AlFe-PILC.

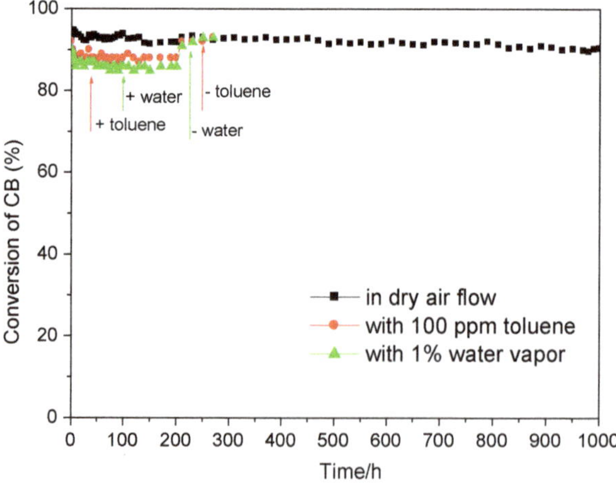

Figure 5. Lifetime test performed for CrCe(5:1)/AlFe-PILC at 280 °C. CB concentration: 500 ppm; gas hourly space velocity (GHSV): 25,000 h^{-1}; catalyst amount: 350 mg.

3.2.2. Effect of CB Concentration and Gas Hourly Space Velocity

Figure 6 presents the effect of CB inlet concentrations on the catalytic performance of CrCe(5:1)/AlFe-PILC. The change of its inlet concentration had a great influence on CB conversion from 500 to 2500 ppm. Furthermore, when the inlet concentration was in the range of 500 to 1500 ppm, CB conversion increased appreciably. This was primarily because low concentration CB only provided a small amount for chemisorbed CB on catalyst active sites and could act as the controlling factor of the reaction. However, as the concentration of CB continued to increase, CB conversion decreased until it was completely prohibited, which may be related to chemisorbed oxygen on the catalyst active sites becoming the reaction controlling factor [34]. The result indicated that CB degradation combustion proceeds via a Langmuir–Hinshelwood (L-H) mechanism, and this catalyst could be used for removing CB waste gases with a wide range of concentrations.

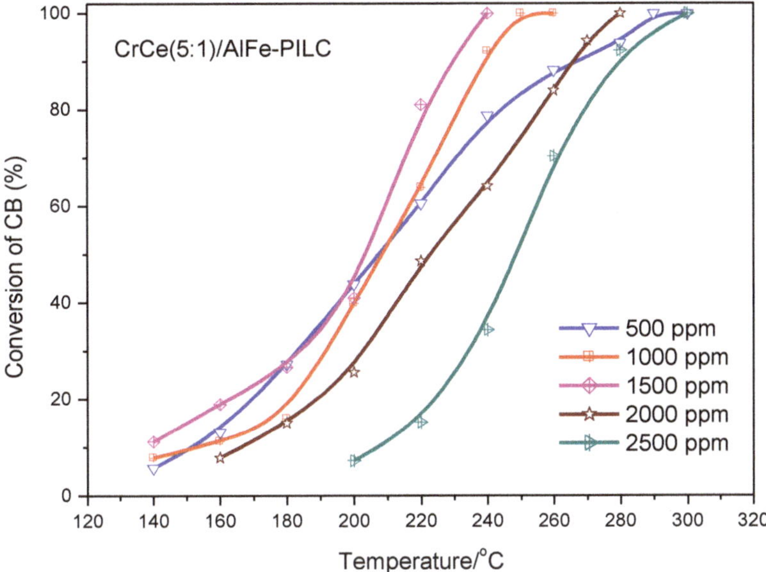

Figure 6. The effect of inlet concentration on CB catalytic combustion over CrCe(5:1)/AlFe-PILC. CB concentration: 500–2500 ppm; GHSV: 25,000 h^{-1}; catalyst amount: 350 mg.

Figure 7 shows the effect of GHSV on the CB catalytic combustion activities over CrCe(5:1)/AlFe-PILC. GHSV is the gas hourly space velocity. To calculate this parameter, the flow rate of feed gas (involved inert and main components) can be adjusted. Then, GHSV is the ratio of gas flow rate in standard conditions to the volume of the catalyst. Increasing GHSV slightly decreased the catalytic performance, indicating that this catalyst was highly effective for CB destruction in different reaction conditions. The catalyst active sites were already fully occupied, even with the lowest GHSV used in this work, and more reactant molecules provided by high GHSV could not be chemisorbed and reacted. Thus, high temperature was required to obtain the same conversion with high GHSV.

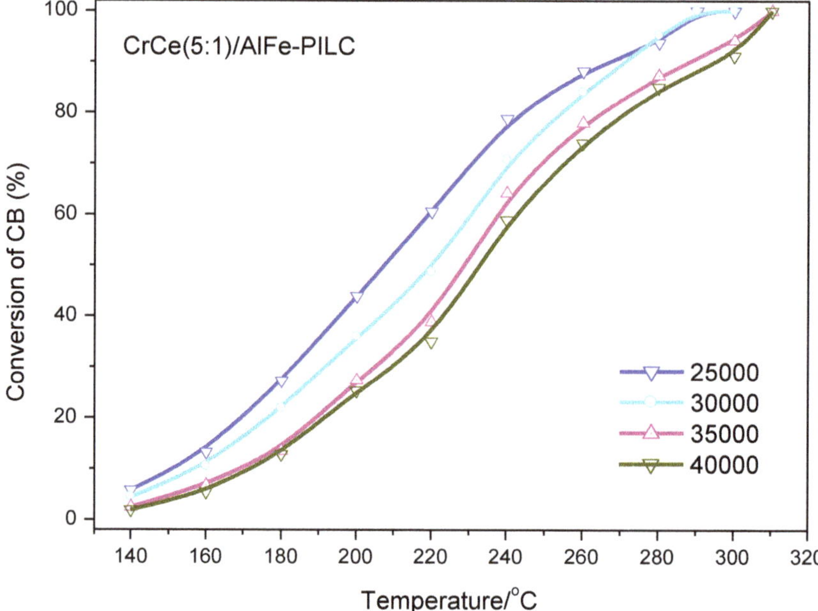

Figure 7. CB conversions vs. temperature over CrCe(5:1)/AlFe-PILC under the conditions of CB concentration at 500 ppm and GHSV at 25,000–40,000 h^{-1}.

3.3. Temperature-Programmed Reaction Studies

3.3.1. H_2-TPR Analysis

H_2-TPR profiles of the catalysts are shown in Figure 8. The reduction of Fe_2O_3 species was obvious in all the catalysts (γ peak), which was from the relatively high contents of Fe_2O_3 (4.45% in the original clay) [35,36]. Compared with the γ peak area from Cr/Na-Mt, the areas from Cr/AlFe-PILC and CrCe(5:1)/AlFe-PILC increased, revealing that more iron oxide species were formed as Fe_2O_3 pillars in AlFe pillaring. In the case of Ce/Na-Mt, it was beneficial for the reduction of surface and bulk CeO_2 to have two reduction peaks at 541 and 745 °C. There were two reduction peaks below 650 °C in the Na-Mt and AlFe-PILC-supported Cr catalysts, which indicated that peaks α_1 and α_2 were the reduction peaks of the surface and inside Cr_2O_3, respectively. For CrCe(5:1)/AlFe-PILC, peaks α_1 and α_2 were divided into two or three peaks, which suggested the better-dispersed Cr_2O_3 on the AlFe-PILC support. Compared with Cr/Na-Mt, the reduction peaks of CrCe(5:1)/AlFe-PILC systematically shifted to lower temperatures, indicating CeO_2 improved the reducibility of Cr_2O_3 by increasing Cr_2O_3 dispersion and lattice oxygen mobility. The peak β_2 of CrCe(5:1)/AlFe-PILC at 588 °C was the reduction peak of bulk CeO_2, and the peak of surface CeO_2 overlapped with peak α_2. The CeO_2 reduction peak was shifted toward lower temperatures compared with that from Ce/Na-Mt. This shift occurred because Cr_2O_3 underwent a stronger oxidation process and could be more easily reduced, allowing it to interact with CeO_2 to produce a reduction peak at a lower temperature. It suggested that the interaction between Cr_2O_3 and CeO_2 species weakened the Ce-O bond and promoted the reduction of CeO_2. The α peak temperatures followed: Cr/Na-Mt > Cr/AlFe-PILC > CrCe(5:1)/AlFe-PILC. The results indicated that the interaction between Cr_2O_3 and CeO_2 species could improve the mobility of oxygen species in the catalysts, thus improving the reduction of both Cr_2O_3 and CeO_2 species.

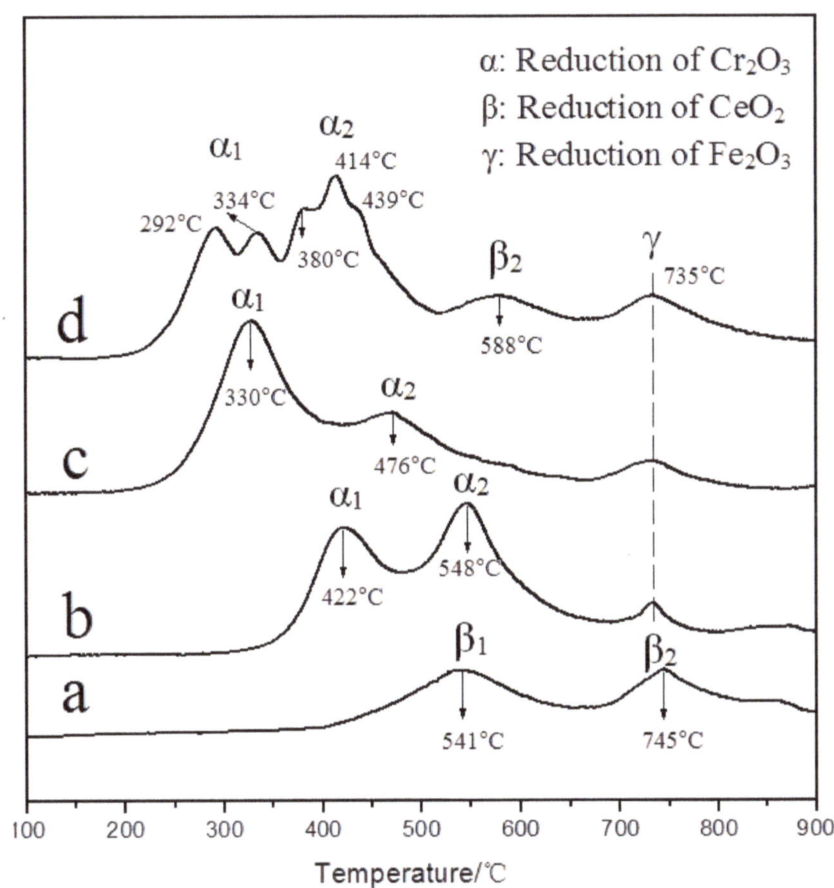

Figure 8. H_2-TPR spectra: (**a**) Ce/Na-Mt, (**b**) Cr/Na-Mt, (**c**) Cr/AlFe-PILC, and (**d**) CrCe(5:1)/AlFe-PILC.

3.3.2. TPD and TPSR Analysis

The adsorption/desorption of CB, catalytic combustion behavior, and the evolution of the main products (CO_2, H_2O, and HCl) over the catalysts were investigated using CB-TPD/TPSR techniques (Figure 9). As it was mentioned previously, the CB combustion on CrCe(5:1)/AlFe-PILC catalyst proceeded via an L-H mechanism, where the adsorption of reactants on the catalyst active sites was a critical step. In Figure 9a, CrCe(5:1)/Na-Mt and CrCe(5:1)/AlFe-PILC showed different CB adsorption capacities. The CB absorption capacities of CrCe(5:1)/AlFe-PILC (44.8 μmol/g) was obviously stronger than CrCe(5:1)/Na-Mt (7.9 μmol/g) by integrating over the absorption spectra. The above results fully proved that clay materials with larger S_{BET}, V_p, and d_{001}-value favor CB adsorption. In Figure 9b, the temperature of CB desorption peaked for CrCe(5:1)/Na-Mt and CrCe(5:1)/AlFe-PILC were 145 °C and 198 °C, respectively, indicating that the interaction of CB and CrCe(5:1)/AlFe-PILC was stronger than with CrCe(5:1)/Na-Mt. Therefore, CB could remain inside the pores or outside the surface of CrCe(5:1)/AlFe-PILC for a longer time, being conducive to the adsorption and catalytic degradation of CB. The results indicated that improved structure and the strong interaction between CrCe mixed oxides with AlFe-PILC enhanced the adsorption of CB.

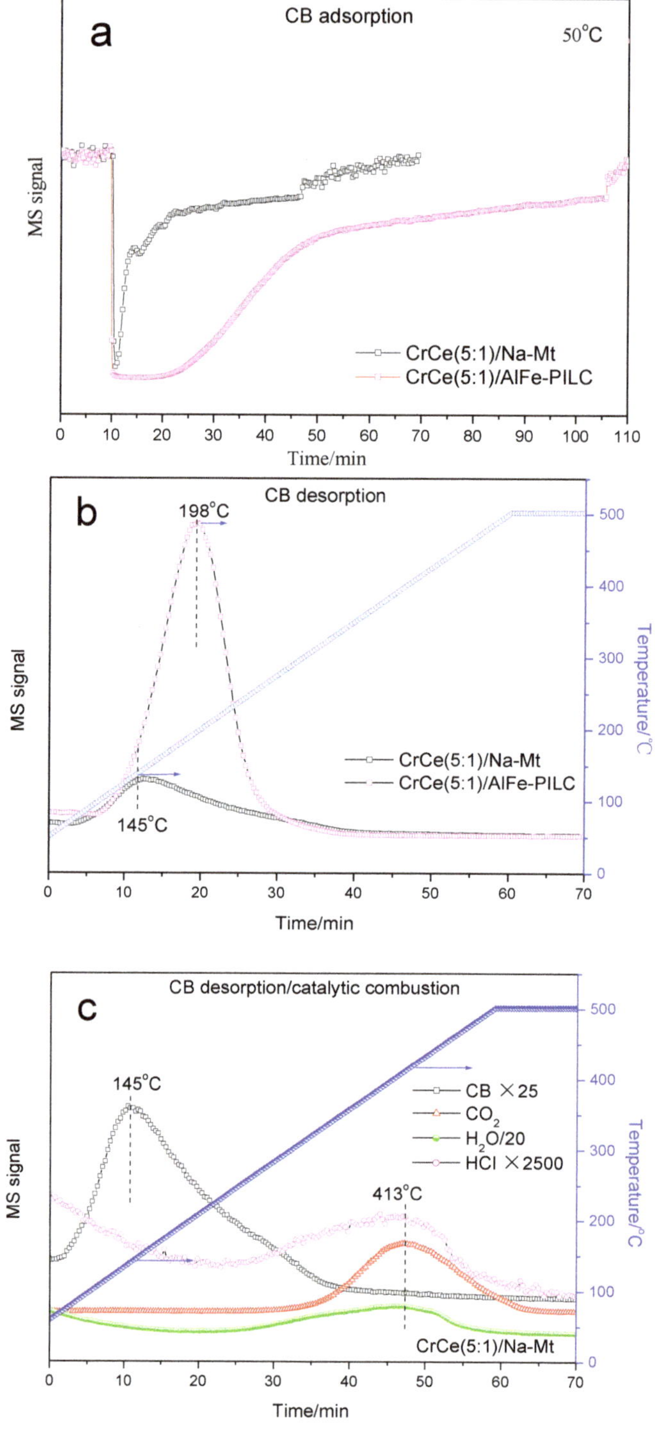

Figure 9. *Cont.*

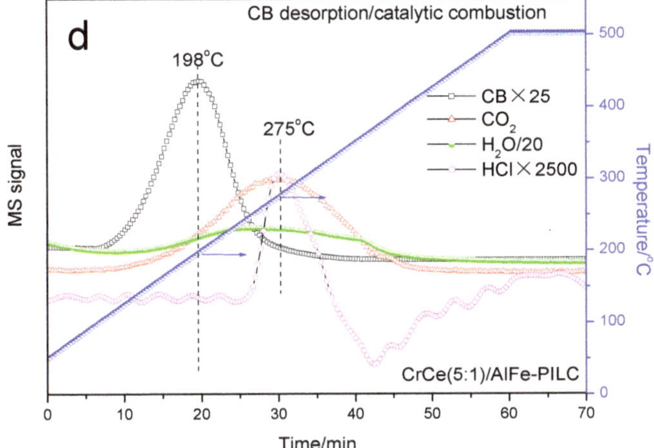

Figure 9. Performance of CB adsorption/desorption and catalytic combustion over CrCe(5:1)/Na-Mt and CrCe(5:1)/AlFe-PILC. (**a**) CB adsorption of CrCe(5:1)/Na-Mt and CrCe(5:1)/AlFe-PILC; (**b**) CB desorption of CrCe(5:1)/Na-Mt and CrCe(5:1)/AlFe-PILC; (**c**) CB desorption/catalytic combustion of CrCe(5:1)/Na-Mt; (**d**) CB desorption/catalytic combustion of CrCe(5:1)/AlFe-PILC.

As shown in Figure 9c,d, CB desorption was accompanied with CB combustion under O_2/Ar, and the adsorbed CB species reacted with lattice O from $CrCeO_x$ to form CO_2, H_2O, and HCl. CB was completely reacted over CrCe(5:1)/AlFe-PILC at about 300 °C, while it needed about 440 °C on CrCe(5:1)/Na-Mt. CO_2, H_2O, and HCl were detected, but CO and Cl_2 were not detected, indicating that the catalysts in the study had high selectivity to HCl and CO_2 formation. It was notable that the peak temperature of the products for CrCe(5:1)/Na-Mt was at 413 °C, which was much higher than that of CB desorption peak temperature (145 °C). However, the peak temperature of product for CrCe(5:1)/AlFe-PILC was at 275 °C, which was close to that of CB desorption (198 °C). This phenomenon can explain why CrCe(5:1)/AlFe-PILC had the highest CB degradation activities compared to other catalysts in this work. The larger overlapped region between CB desorption and catalytic combustion, the better the catalytic performance. Therefore, tuning the CB adsorption and catalytic properties was a key to designing an efficient catalyst for CB catalytic combustion.

In order to find out the relationship between the oxygen species absorbed on the catalyst surface and the catalytic properties, O_2-TPD were investigated from 50 to 900 °C. The O_2-TPD plots for Cr and CrCe metal oxide catalysts consisted of oxygen desorption regions shown in Figure 10. There were three types of desorption peaks, the α desorption peak, the β desorption peak, and the γ desorption peak. Furthermore, these three peaks could be assigned to superoxide ion O_2^-, peroxide ion O_2^{2-}/O^-, and lattice oxygen ion O^{2-}, respectively [37,38]. Increasing the temperature is beneficial to increase the rate of desorption and transformation of superoxide species into O_2^{2-}, O^-, and $O_{lattice}^{2-}$ [39]. It can be seen that the α and β desorption peaks follow: CrCe(5:1)/AlFe-PILC < Cr/AlFe-PILC < CrCe(5:1)/Na-Mt < Cr/Na-Mt, which was in good agreement with the aforementioned catalytic performance of CB combustion. It was worth mentioning that the total amount of surface-active oxygen species, in terms of the sum of α and β desorption areas, follows the same sequence of the peaks. It can be observed from the γ desorption peak that adding CeO_2 increased the desorption area of lattice oxygen ion O^{2-} compared with the non-doped catalyst. Thus CrCe(5:1)/AlFe-PILC exhibited the highest oxidation performance since electrophilic O_{ads} (O_2^-, O_2^{2-}, O^-) played a critical role in the complete oxidation of organic compounds [40].

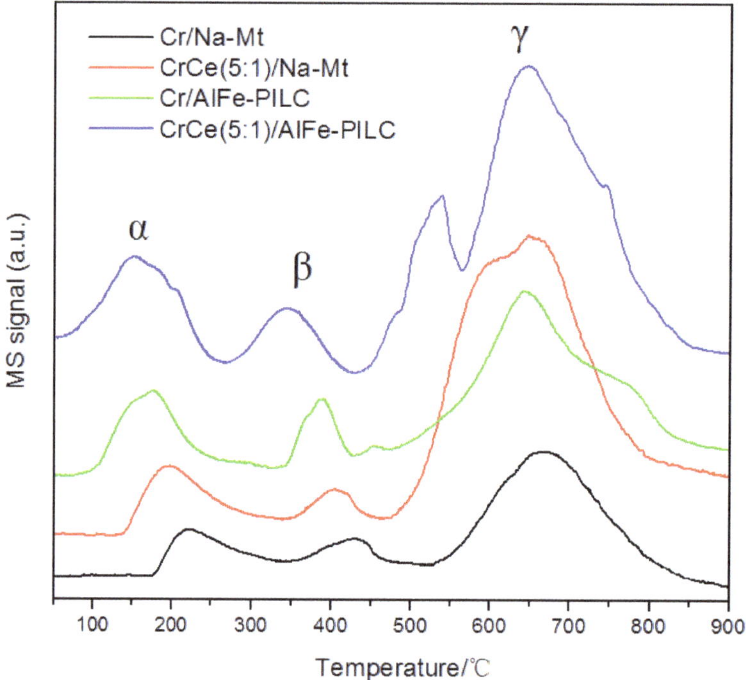

Figure 10. O_2-TPD profiles of Cr/Na-Mt, CrCe(5:1)/Na-Mt, Cr/AlFe-PILC, and CrCe(5:1)/AlFe-PILC.

4. Conclusions

In this paper, AlFe-PILC supported CrCe mixed oxides are synthesized and used for adsorption/desorption and catalytic combustion of CB. A series of characterization methods were used to investigate the structure and redox properties of these materials, including HRTEM-EDS, H_2-TPR, TPD/TPSR, and O_2-TPD. Comparing the results of the S_{BET}, V_p, and d_{001}-value, AlFe-PILC performed better than Na-Mt. Without doubt, AlFe-PILC synthesized in this study constituted a class of porous materials with excellent properties. A large number of micro-mesoporosity and Ce was added to the AlFe-PILC to optimize its structure and improve the dispersion of Cr_2O_3 particles on the AlFe-PILC. XRD analysis and HRTEM images clearly revealed the stable layered structure with the d_{001} value ≈1.91 nm and well-dispersed active species in AlFe-PILC. The addition of Ce and optimized structure of support greatly improved the oxidative property of Cr_2O_3. CB-TPD experiments reveal that the optimized structure coupled with the strong interaction between CrCe metal oxides and AlFe-PILC enhanced CB adsorption capacity and adsorption strength. CB-TPSR results showed that the larger the overlapped region between CB desorption and the catalytic combustion, the better the catalytic performance. In particular, CrCe(5:1)/AlFe-PILC show an excellent catalytic property, and stability was due to the lower temperature of completely degraded CB (approximately 290 °C) and the conversion remained stable for 1000 h. CB catalytic combustion on CrCe/AlFe-PILC catalyst was via a Langmuir–Hinshelwood mechanism, and adjusting adsorption/desorption properties was one of the most important factors for designing efficient catalysts. CrCe/AlFe-PILC also exhibited good durability for CB destruction, both in the humid condition and in the presence of toluene; therefore, this catalyst deserves wide attention and it is a potential prospect for industrial application.

Author Contributions: S.Z. designed the experiments; Y.Q. and D.S. performed the experiments; S.Z. and X.W. wrote the paper; N.Y. provided some experimental equipment and guided the experiments.

Funding: Zhejiang Public Welfare Technology Research Project (LGG19B070003), the Foundation of Science and Technology of the Shaoxing City (2018C10019) and 2018 Zhejiang Province Innovation Training Program for College Students (2018R432031).

Acknowledgments: We would like to acknowledge the financial support from Innovation Team of Huzhou South Taihu Elite Program. Moreover, we would also grateful to Zhejiang Da-Feng Automobile Technology Co., Ltd for the related experiment and test.

Conflicts of Interest: The authors declare no conflict of interest.

References

1. De Rivas, B.; Guillén-Hurtado, N.; López-Fonseca, R.; Coloma-Pascual, F.; García-García, A.; Gutiérrez-Ortiz, J.I.; Bueno-López, A. Activity, selectivity and stability of praseodymium-doped CeO_2 for chlorinated VOCs catalytic combustion. *Appl. Catal. B Environ.* **2012**, *121*, 162–170. [CrossRef]
2. Weng, X.L.; Sun, P.F.; Long, Y.; Meng, Q.J.; Wu, Z.B. Catalytic Oxidation of chlorobenzene over $Mn_xCe_{1-x}O_2$/HZSM-5 catalysts: A study with practical implications. *Environ. Sci. Technol.* **2017**, *51*, 8057–8066. [CrossRef] [PubMed]
3. Huang, B.B.; Lei, C.; Wei, C.H.; Zeng, G.M. Chlorinated volatile organic compounds (Cl-VOCs) in environment—Sources, potential human health impacts, and current remediation technologies. *Environ. Int.* **2014**, *71*, 118–138. [CrossRef] [PubMed]
4. Giraudon, J.M.; Nguyen, T.B.; Leclercq, G.; Siffert, S.; Lamonier, J.F.; Aboukais, A.; Vantomme, A.; Su, B.L. Chlorobenzene total oxidation over palladium supported on ZrO_2, TiO_2 nanostructured supports. *Catal. Today* **2008**, *137*, 379–384. [CrossRef]
5. Cai, T.; Huang, H.; Deng, W.; Dai, Q.G.; Liu, W.; Wang, X.Y. Catalytic combustion of 1,2-dichlorobenzene at low temperature over Mn-modified Co_3O_4 catalysts. *Appl. Catal. B Environ.* **2015**, *166*, 393–405. [CrossRef]
6. Cao, S.; Fei, X.Q.; Wen, Y.X.; Sun, Z.X.; Wang, H.Q.; Wu, Z.B. Bimodal mesoporous TiO_2 supported Pt, Pd and Ru catalysts and their catalytic performance and deactivation mechanism for catalytic combustion of Dichloromethane (CH_2Cl_2). *Appl. Catal. A Gen.* **2018**, *550*, 20–27. [CrossRef]
7. Huang, H.; Gu, Y.F.; Zhao, J.; Wang, X.Y. Catalytic combustion of chlorobenzene over VOx/CeO_2 catalysts. *J. Catal.* **2015**, *326*, 54–68. [CrossRef]
8. He, F.; Luo, J.Q.; Liu, S.T. Novel metal loaded KIT-6 catalysts and their applications in the catalytic combustion of chlorobenzene. *Chem. Eng. J.* **2016**, *294*, 362–370. [CrossRef]
9. Huang, H.; Dai, Q.G.; Wang, X.Y. Morphology effect of Ru/CeO_2 catalysts for the catalytic combustion of chlorobenzene. *Appl. Catal. B Environ.* **2014**, *158*, 96–105. [CrossRef]
10. Gannoun, C.; Turki, A.; Kochkar, H.; Delaigle, R.; Eloy, P.; Ghorbel, A.; Gaigneaux, E.M. Elaboration and characterization of sulfated and unsulfated V_2O_5/TiO_2 nanotubes catalysts for chlorobenzene total oxidation. *Appl. Catal. B Environ.* **2014**, *147*, 58–64. [CrossRef]
11. Scirè, S.; Minicò, S.; Crisafulli, C. Pt catalysts supported on H-type zeolites for the catalytic combustion of chlorobenzene. *Appl. Catal. B Environ.* **2003**, *45*, 117–125. [CrossRef]
12. Zhao, W.; Cheng, J.; Wang, L.; Chu, J.L.; Qu, J.K.; Liu, Y.H.; Li, S.H.; Zhang, H.; Wang, J.C.; Hao, Z.P.; Qi, T. Catalytic combustion of chlorobenzene on the Ln modified Co/HMS. *Appl. Catal. B Environ.* **2012**, *127*, 246–254. [CrossRef]
13. Shao, Y.; Xu, Z.Y.; Wan, H.Q.; Chen, H.; Liu, F.L.; Li, L.Y.; Zheng, S.R. Influence of ZrO_2 properties on catalytic hydrodechlorination of chlorobenzene over Pd/ZrO_2 catalysts. *J. Hazard. Mater.* **2010**, *179*, 135–140. [CrossRef] [PubMed]
14. Dai, Q.G.; Bai, S.X.; Wang, J.W.; Li, M.; Wang, X.Y.; Lu, G.Z. The effect of TiO_2 doping on catalytic performances of Ru/CeO_2 catalysts during catalytic combustion of chlorobenzene. *Appl. Catal. B Environ.* **2013**, *142*, 222–233. [CrossRef]
15. Gu, Y.L.; Yang, Y.X.; Qiu, Y.M.; Sun, K.P.; Xu, X.L. Combustion of dichloromethane using copper–manganese oxides supported on zirconium modified titanium-aluminum catalysts. *Catal. Commun.* **2010**, *12*, 277–281. [CrossRef]

16. Sun, P.F.; Wang, W.L.; Dai, X.X.; Weng, X.L.; Wu, Z.B. Mechanism study on catalytic oxidation of chlorobenzene over $Mn_xCe_{1-x}O_2$/H-ZSM5 catalysts under dry and humid conditions. *Appl. Catal. B Environ.* **2016**, *198*, 389–397. [CrossRef]
17. He, C.; Xu, B.T.; Shi, J.W.; Qiao, N.L.; Hao, Z.P.; Zhao, J.L. Catalytic destruction of chlorobenzene over mesoporous $ACeO_x$ (A = Co, Cu, Fe, Mn, or Zr) composites prepared by inorganic metal precursor spontaneous precipitation. *Fuel Process. Technol.* **2015**, *130*, 179–187. [CrossRef]
18. Debecker, D.P.; Bertinchamps, F.; Blangenois, N.; Eloy, P.; Gaigneaux, E.M. On the impact of the choice of model VOC in the evaluation of V-based catalysts for the total oxidation of dioxins: Furan vs. chlorobenzene. *Appl. Catal. B Environ.* **2007**, *74*, 223–232. [CrossRef]
19. Hetrick, C.E.; Lichtenberger, J.; Amiridis, M.D. Catalytic oxidation of chlorophenol over V_2O_5/TiO_2 catalysts. *Appl. Catal. B Environ.* **2008**, *77*, 255–263. [CrossRef]
20. Wang, J.; Wang, X.; Liu, X.L.; Zeng, J.L.; Guo, Y.Y.; Zhu, T.Y. Kinetics and mechanism study on catalytic oxidation of chlorobenzene over V_2O_5/TiO_2 catalysts. *J. Mol. Catal. A Chem.* **2015**, *402*, 1–9. [CrossRef]
21. Choi, K.H.; Shokouhimehr, M.; Sung, Y.E. Heterogeneous Suzuki cross-coupling reaction catalyzed by magnetically recyclable nanocatalyst. *Bull. Korean Chem. Soc.* **2013**, *34*, 1477–1480. [CrossRef]
22. Shokouhimehr, M.; Asl, M.S.; Mazinani, B. Modulated large-pore mesoporous silica as an efficient base catalyst for the Henry reaction. *Res. Chem. Intermed.* **2018**, *44*, 1617–1626. [CrossRef]
23. Shokouhimehr, M.; Hong, K.; Lee, T.H.; Moon, C.W.; Hong, S.P.; Zhang, K.Q.; Suh, J.M.; Choi, K.S.; Varma, R.S.; Jang, H.W. Magnetically retrievable nanocomposite adorned with Pd nanocatalysts: Efficient reduction of nitroaromatics in aqueous media. *Green Chem.* **2018**, *20*, 3809–3817. [CrossRef]
24. Zhang, K.Q.; Suh, J.M.; Choi, J.W.; Jang, H.W.; Shokouhimehr, M.; Varma, R.S. Recent advances in the nanocatalyst-assisted $NaBH_4$ reduction of nitroaromatics in Water. *ACS Omega* **2019**, *4*, 483–495. [CrossRef]
25. Shokouhimehr, M. Magnetically separable and sustainable nanostructured catalysts for heterogeneous reduction of nitroaromatics. *Catalysts* **2015**, *5*, 534–560. [CrossRef]
26. Li, D.; Li, C.S.; Suzuki, K.Z. Catalytic oxidation of VOCs over Al- and Fe-pillared montmorillonite. *Appl. Clay Sci.* **2013**, *77*, 56–60. [CrossRef]
27. Aznárez, A.; Korili, S.A.; Gil, A. The promoting effect of cerium on the characteristics and catalytic performance of palladium supported on alumina pillared clays for the combustion of propene. *Appl. Catal. A Gen.* **2014**, *474*, 95–99. [CrossRef]
28. Aznárez, A.; Delaigle, R.; Eloy, P.; Gaigneaux, E.M.; Korili, S.A.; Gil, A. Catalysts based on pillared clays for the oxidation of chlorobenzene. *Catal. Today* **2015**, *246*, 15–27. [CrossRef]
29. Vicente, M.A.; Gil, A.; Bergaya, F. Pillared Clays and Clay Minerals. In *Developments in Clay Science 5A*; Elsevier Science: Amsterdam, The Netherlands, 2013; Chapter 10.5; pp. 523–557.
30. Zuo, S.F.; Yang, P.; Wang, X.Q. Efficient and environmentally friendly synthesis of AlFe-PILC-supported MnCe catalysts for benzene combustion. *ACS Omega* **2017**, *2*, 5179–5186. [CrossRef]
31. Zuo, S.F.; Ding, M.L.; Tong, J.; Feng, L.C.; Qi, C.Z. Study on the preparation and characterization of a titanium-pillared clay-supported CrCe catalyst and its application to the degradation of a low concentration of chlorobenzene. *Appl. Clay Sci.* **2015**, *105*, 118–123. [CrossRef]
32. Cheng, Z.; Chen, Z.; Li, J.R.; Zuo, S.F.; Yang, P. Mesoporous silica-pillared clays supported nanosized Co_3O_4-CeO_2 for catalytic combustion of toluene. *Appl. Surf. Sci.* **2018**, *459*, 32–39. [CrossRef]
33. Booij, E.; Kloprogge, J.T.; van Veen, J.R. Large pore REE/Al pillared bentonites: Preparation, structural aspects and catalytic properties. *Appl. Clay Sci.* **1996**, *11*, 155–162. [CrossRef]
34. He, C.; Yu, Y.K.; Shen, Q.; Chen, J.S.; Qiao, N.L. Catalytic behavior and synergistic effect of nanostructured mesoporous CuO-MnO_x-CeO_2 catalysts for chlorobenzene destruction. *Appl. Surf. Sci.* **2014**, *297*, 59–69. [CrossRef]
35. Basińska, A.; Jóźwiak, W.K.; Góralski, J.; Domka, F. The behaviour of Ru/Fe_2O_3 catalysts and Fe_2O_3 supports in the TPR and TPO conditions. *Appl. Catal. A Gen.* **2000**, *190*, 107–115. [CrossRef]
36. Durán, F.G.; Barbero, B.P.; Cadús, L.E.; Rojas, C.; Centeno, M.A.; Odriozola, J.A. Manganese and iron oxides as combustion catalysts of volatile organic compounds. *Appl. Catal. B Environ.* **2009**, *92*, 194–201. [CrossRef]
37. Zhao, B.; Wang, R.J.; Yang, X.X. Simultaneous catalytic removal of NO_x and diesel soot particulates over La1-x$Ce_x$$NiO_3$ perovskite oxide catalysts. *Catal. Commun.* **2009**, *10*, 1029–1033. [CrossRef]
38. Wang, X.Y.; Zuo, J.C.; Luo, Y.J.; Jiang, L.L. New route to $CeO_2/LaCoO_3$ with high oxygen mobility for total benzene oxidation. *Appl. Surf. Sci.* **2017**, *396*, 95–101. [CrossRef]

39. Lambrou, P.S.; Efstathiou, A.M. The effects of Fe on the oxygen storage and release properties of model Pd–Rh/CeO_2–Al_2O_3 three-way catalyst. *J. Catal.* **2006**, *240*, 182–193. [CrossRef]
40. Feng, Y.J.; Li, L.; Niu, S.F.; Qu, Y.; Zhang, Q.; Li, Y.S.; Zhao, W.R.; Li, H.; Shi, J.L. Controlled synthesis of highly active mesoporous Co_3O_4 polycrystals for low temperature CO oxidation. *Appl. Catal. B Environ.* **2012**, *111*, 461–466. [CrossRef]

Article

Combined Effect of Pressure and Carbon Dioxide Activation on Porous Structure of Lignite Chars

Natalia Howaniec

Department of Energy Saving and Air Protection, Central Mining Institute, Pl. Gwarkow 1, 40-166 Katowice, Poland; n.howaniec@gig.eu; Tel.: +48-32-259-2219

Received: 3 April 2019; Accepted: 22 April 2019; Published: 23 April 2019

Abstract: Lignite is an important natural resource with the application potential covering present and future energy systems, including conventional power plants and gasification systems. Lignite is also a valuable precursor for the production of porous materials of tailored properties for various environmental applications, including the removal of contaminants from gaseous or liquid media. Although the lignite-based activated carbons are commercially available, various approaches to produce carbon materials of desired properties are still being reported, covering temperature, partial oxidation and chemical activation effects on surface and structural properties of these materials. Limited data is, however, available on the effects of pressure as the activation parameter in shaping the porous structure of carbonaceous materials, in particularly lignite-derived. In the study presented the combined effect of carbon dioxide activation and pressure in the range of 1–3 MPa at the temperature of 800 °C on the development of porous structure of lignite chars was reported. The study was also focused on poor-quality resources valorization by using a relatively low calorific value, low volatiles and high ash content lignite as a carbon material precursor. The results showed that the application of pressure in carbon dioxide-activation process at 800 °C results in generation of chars of comparable or higher specific surface area than the carbon materials previously received with demineralization and carbon dioxide activation of lignite. They also proved that the combined pressure and carbon dioxide activation may be effectively applied in conversion of low quality lignite into valuable porous materials.

Keywords: lignite; porous structure; carbon dioxide; activation; pressure

1. Introduction

Lignite is a considerable element of world coal resources [1]. As such, it is not only an important energy resource applied currently in conventional power plants but also a fuel particularly suitable for gasification and co-gasification systems for its higher reactivity when compared to bituminous coals [2–4]. Lignite is also a valuable parent material for the production of porous materials of tailored properties for various industrial applications [5–8]. The majority of them concerns sorption processes in the removal of contaminants, e.g., phenol, mercury, sulphur oxides, copper, and organic compounds from gaseous or liquid media [9–14]. Although the lignite-based activated carbons are commercially available, various approaches to produce carbon materials of desired properties are still being reported. The vast literature is available on the application of the temperature and various oxidizing agents [5,7,11–17] as well as acidic or basic treatment [7,10,18–20] in shaping the surface and structural properties of lignite-derived carbon materials. Limited data is, however, available on the effects of pressure as the activation parameter in shaping the porous structure of carbonaceous materials [21–23], in particularly lignite [24–26]. A few studies considering the development of porous structure of bituminous coals under carbon dioxide atmosphere [27,28] or inert gas [24,29,30] and elevated pressure are available. Porous structure development of chars is also considered in the literature in terms of lignite suitability and chars reactivity in gasification process, in particularly with the incorporation of carbon dioxide in a valorization cycle as a gasification process reactant [20,31–34]. Previous studies showed that the values of surface area of bituminous coal chars developed under carbon dioxide atmosphere in general drop with process pressure [27,28]. Swelling properties have been reported to influence the porous structure of coal chars under inert gas atmosphere and pressurized conditions, and although this effect is enhanced with increased coal rank, it also depends on volatiles and specific petrographic components content [24,29] and vary with pressure values [5,30]. The amounts of particular mineral components, differing in rates of expansion, e.g., kaolinite, quartz, pyrite, and calcite affect the porous structure development of lignite chars at increased temperature. The mineral matter in bituminous coal-derived chars was also reported to be severely affected qualitatively and quantitatively when treated with carbon dioxide at 900 °C. The inorganic amorphous phase was decomposed and the reduced mineral forms were oxidized and reacted with aluminosilicates forming calcium and iron minerals [35].

The combined effect of carbon dioxide activation and pressure on the development of porous structure of lignite chars has not been reported so far. Therefore, the experimental study on the application of pressure in the range of 1–3 MPa in carbon dioxide-activation of lignite chars was performed and its results are presented in this paper. The lignite of relatively low calorific value and high ash content was selected as a precursor which makes the study valid also in the context of poor quality resources valorization.

2. Materials and Methods

The lignite of relatively low calorific value, high ash and sulfur content, provided by Polish opencast mine from Szczercow deposit, was selected as the chars precursor. Lignite of Szczercow deposit is characterized by high huminite maceral group content, of approx. 82% vol., including densinite share of 44% vol. and atrinite content of 23% vol. The liptinite and intertinite contents are of 7 and 4% vol. Highly porous macerals, like textinite, ulminite and atrinite amount in total to approximately 37% vol. [33]. The mineral matter content is approximately 8% vol. with the dominance of clay minerals. The proximate and ultimate analyses of lignite tested were performed in an accredited laboratory in compliance with the relevant standards and are given in Table 1.

Table 1. Physical and chemical properties of lignite sample tested.

Parameter, Unit	Value
Proximate analysis	-
Total moisture [1], % *w/w*	12.97
Ash [1], % *w/w*	23.07
Volatiles [1], % *w/w*	37.53
Fixed carbon [2], % *w/w*	26.43
Ultimate analysis	-
Sulfur [3], % *w/w*	2.84
Carbon [4], % *w/w*	42.63
Hydrogen [4], % *w/w*	3.21
Nitrogen [4], % *w/w*	0.34
Oxygen [2], % *w/w*	15.37
Heating value	-
Higher heating value [5], kJ/kg	16,602
Lower heating value [5], kJ/kg	15,585

[1] PN-G-04560:1998 with the use of automatic thermogravimetric analyzers LECO (St. Joseph, MI, USA): TGA 701 and MAC 500; [2] PN-G-04516:1998 calculated by difference; [3] PN-G-0484:2001 with the use of an automatic analyzer TruSpec S by LECO; [4] PN-G-04571:1998 with the use of an automatic analyzer TruSpec CHN by LECO; [5] PN-G-04513:1981 with the use of LECO calorimeters: AC-600 and AC-350.

Lignite sample of 1 g was heated in a high-pressure thermogravimetric analyzer (Rubotherm GmbH, Bochum, Germany) with the heating rate of 20 °C/min in an argon atmosphere to the final process temperature of 800 °C (see Figure 1) and pressurized to the final process pressure of 1, 2 or 3 MPa, respectively. When the final process temperature and pressure were reached carbon dioxide was introduced to the reactor with a flow rate of 100 mL/min for 120 min.

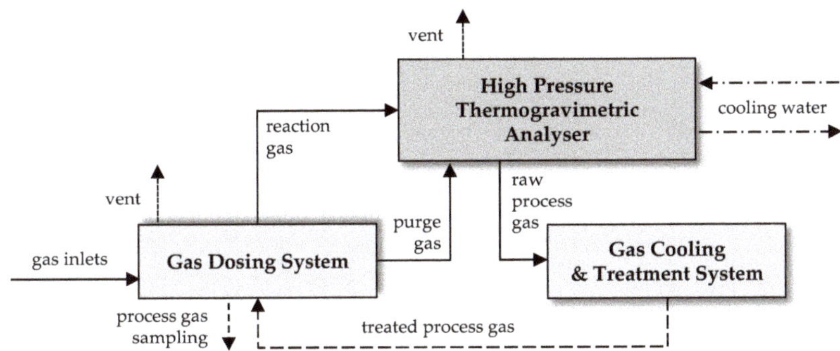

Figure 1. Schematic diagram of the system for carbon materials preparation at high temperature and pressure and with the use of carbon dioxide activation.

The resulting carbon materials were outgassed at 120 °C overnight and analyzed in terms of their porous structure parameters with the application of a gas sorption analyzer Autosorb iQ (Quantachrome Instruments, Boynton Beach, FL, USA). Based on the nitrogen sorption isotherm data acquired at −196 °C, the specific surface area and pore size distribution were determined with the application of the multi-point BET method [36] and the Density Functional Theory (DFT) [37], respectively. The total pore volume was quantified as the volume at the relative pressure of 0.99. The narrow micropore area and volume were further analyzed on the basis of the carbon dioxide isotherm at 0 °C and the Monte Carlo (MC) method [38]. The surface properties of the resulting carbon materials were also explored with the use of a scanning electron microscope SU-3500N (Hitachi High-technologies Corporation, Tokyo, Japan).

3. Results and Discussion

The porous structure of carbon materials produced from low quality lignite at the temperature of 800 °C and under the pressure of 1–3 MPa with a carbon dioxide activation step was complex and composed of micro- and mesopores. High uptake at low relative pressures, which may be seen in Figure 2, presenting the exemplary nitrogen isotherm for lignite chars tested, is indicative of micropores present in the porous structure of lignite chars. The occurrence of a hysteresis loop proves that the material is also rich in mesopores and its profile reveals the irregular, slit-like shape of pores [39].

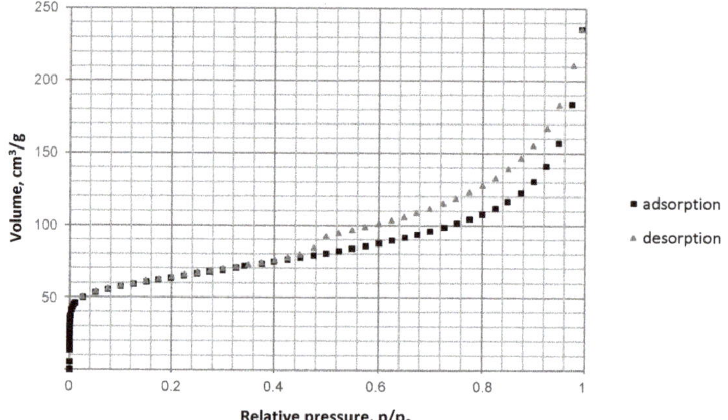

Figure 2. Nitrogen isotherm (−196 °C) for lignite chars produced at 800 °C, under 2 MPa and with carbon dioxide activation.

The results of the porous structure parameters of carbon materials produced under the pressure of 1–3 MPa and at atmospheric pressure, for comparison purposes, are given in Table 2. It may be seen that the rise in pressure resulted in an increased specific surface area values. The highest increase, of 9%, was observed with the change in the process pressure from the atmospheric to 1 MPa. Further increase in pressure in the range 1–3 MPa gave the rise in the specific surface area of 6–7% per 1 MPa.

Table 2. Properties of porous structure of lignite chars determined with the use of nitrogen sorption isotherm at −196 °C and carbon dioxide sorption isotherm at 0 °C.

Sample No.	Pressure under CO_2 Atmosphere, MPa	Multi-Point BET, m^2/g	Average Pore Diameter, nm	Total Pore Volume, cm^3/g	MC Volume, cm^3/g	MC Area, m^2/g
1	0.1	197	7.46	0.368	0.058	177
2	1	215	6.81	0.366	0.065	206
3	2	227	6.41	0.365	0.066	207
4	3	244	6.57	0.400	0.068	212

The average pore diameter showed a decrease with pressure in the range 0.1–2 MPa (see Table 2). The difference in values of the average pore size of carbon materials produced under 2 and 3 MPa was within the experimental error. It implies that the pressure-enhanced development of smaller pores resulting from the devolatilization, moisture release and partial oxidation of carbon with carbon dioxide under the pressure of up to 2 MPa was counteracted with merging of pores in larger structures under the pressure of 3 MPa. A similar trend was also observed previously for lignite [25,26] and bituminous coal [22,26], as well as biomass chars [21,26] with no carbon dioxide activation, although the limiting values of pressure varied between 2 and 3 MPa for various parent materials. The differences in the values of the total pore volume for chars activated with carbon dioxide under the pressure of 0.1–2 MPa were within the experimental error. However, under the highest pressure tested, of 3 MPa, an increase of approximately 8% in the total pore volume was observed (see Table 2).

The pore size distribution (PSD) data showed an increase in a pore volume with pressure applied mainly because of increasing volume of mesopores of a diameter over 5 nm in lignite chars with carbonization pressure from 0.1 to 3 MPa (see Figure 3a). The pore volume of small micropores (diameter below 1 nm) determined based on the nitrogen isotherm increased with a change in the pressure from atmospheric to elevated, and was comparable for chars generated under 1–3 MPa. The variation in process pressure seemed to have no measurable effect on the development of 1–2 nm micropore volume with the carbon dioxide activation at 800 °C, which may be related to closure of these micropores or their merging in larger structures at higher pressures. The latter seems to be also demonstrated by the increasing share of mesopores of a diameter over 5 nm. These pores had a dominant role in the increase in the total DFT pore volume with pressure which amounted to 4, 11 and 17% with pressure rise from atmospheric to 1, 2, and 3 MPa, respectively. The dominant share of micropores in shaping the pore area is also visible (see Figure 3b) as well as the positive effect of pressure applied on the development of the smallest pores (of a diameter below 1 nm) and the total DFT area of pores (Figure 3b). Under 3 MPa the share of mesopores of a diameter over 2 nm in the pore area also slightly increased which again may be indicative of merging of pores in larger structures under the highest pressure tested. The total DFT area of pores increased with change of the process pressure from atmospheric to 1, 2 and 3 MPa of 15%, 18% and 24%, respectively.

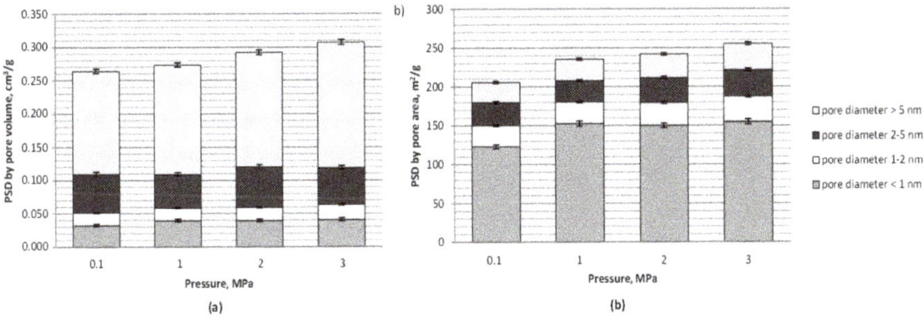

Figure 3. Distribution of: (**a**) pore volume and (**b**) area based on DFT method and nitrogen isotherm (−196 °C) for lignite chars produced at 800 °C, under 0.1–3 MPa and with carbon dioxide activation.

The area and volume of narrow micropores (diameter range of 0.45–1.5 nm), determined based on the carbon dioxide sorption isotherm, showed a clear increase with a change in process conditions from atmospheric to pressurized (Table 2). This was caused mainly by an enhanced development of pores of a diameter 0.65–0.85 nm (see Figure 4). No meaningful difference was however observed with further changes in pressure from 1 to 3 MPa in terms of narrow microporosity, except for a slight increase in the volume of pores of a diameter in the range of 0.85–1.05 nm under 3 MPa, giving a slight rise to the total MC pore volume, which may be the effect of structural rearrangements under the maximum pressure tested.

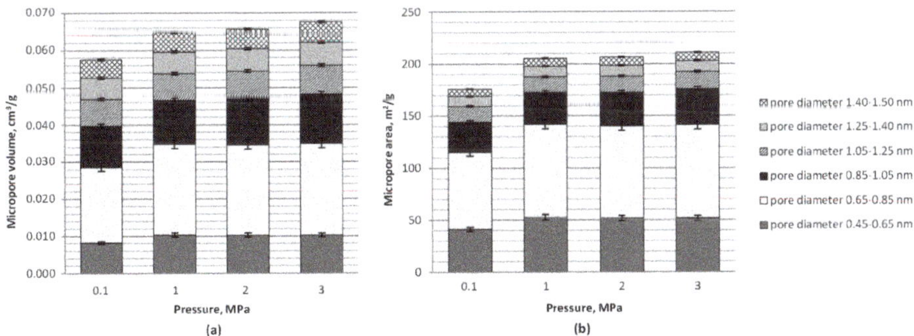

Figure 4. Narrow microporosity determined based on the carbon dioxide isotherm (0 °C) for lignite chars produced at 800 °C, under 0.1–3 MPa and with carbon dioxide activation: (**a**) Micropore volume, and (**b**) micropore area.

The activation with carbon dioxide under the increased pressure seems to be resulting in an increased average pore diameter and the total pore volume, as well as a slightly lower specific surface area when compared to values observed previously for lignite-derived materials with pressure, as the only activation agent in carbonization step at 1000 °C [25] or carbon dioxide as the only activation agent at 900 °C [8]. These effects are clearly also related to the temperature and composition of a parent material since in the studies with CO_2-only activation at 750–900 °C [5] similar values, of 0.336–0.365 cm^3/g to the reported in this study were observed, but for a lignite of a considerable higher volatiles content (48%) and significantly lower ash content (5%), which made it more suitable as a porous material precursor. This proves that the increased pressure and high temperature may be successfully applied instead of partial CO_2-oxidation for production of lignite-derived chars of well-developed surface area and relatively low average pore diameter.

The application of pressure in carbon dioxide activation process at 800 °C enabled production of chars of only slightly lower specific surface area than the ones received with demineralization and carbon dioxide activation step of lignite of a considerably lower ash content (8–9%) than in the study presented here, which implies that pressure may be also considered as an alternative to chemical activation [20]. The total pore volume reported for lignite chars produced with pressure and carbon dioxide activation within the study presented here doubled the values observed for lignite chars of the volatiles content as high as 50% at 800 °C, with chemical demineralization and carbon dioxide activation [11].

These results show that the combined pressure and carbon dioxide activation may be effectively applied in utilization of poor quality parent materials (low volatiles, high ash content) for porous materials development. They also prove that the combination of pressure and carbon dioxide activation results in the development of porous structure of lignite-derived materials of a comparable specific surface area and of a similar or higher total pore volume than observed for complex demineralization, chemical activation and carbon dioxide treatment of the respective parent materials.

As it can be seen from the SEM (define) images presented in Figure 5, the surface of char particles generated under 0.1 and 1 MPa was visibly smoother (Figure 5a,b) than of chars produced under the elevated pressure of 2 and 3 MPa (Figure 5c,d). However, even for the chars generated under 1 MPa a clear difference may be noticed, consisting in a less dense texture and a more complex structure with some cracks and roughness, when compared to chars developed under atmospheric pressure, resulting from volatiles release and lignite carbonization under carbon dioxide atmosphere. This is in line with the variations in the average pore diameter of chars as described above, as well as increased specific surface area and micropore area with pressure applied (Table 2). The chars developed under 2 and 3 MPa (Figure 5c,d, left) have a more pumice-like structure than a plate-like structure characteristic for chars developed under atmospheric pressure and 1 MPa (Figure 5a,b, left). The chars produced under 2 and 3 MPa (Figure 5c,d, right) showed also visibly more expanded cavities and larger cracks than chars generated under lower pressures. They are likely to be composed of a considerable amount of micropores and smaller mesopores of irregular shape as demonstrated also by the nitrogen isotherms shape and DFT data (Figure 2). This means that the elevated pressure resulted in an enhanced porosity of chars under the experimental conditions applied, though there may also have occurred some rearrangements in the porous structure resulting from thermal annealing observed previously for carbon dioxide treatment of bituminous coal chars at the temperature of 800–900 °C [34].

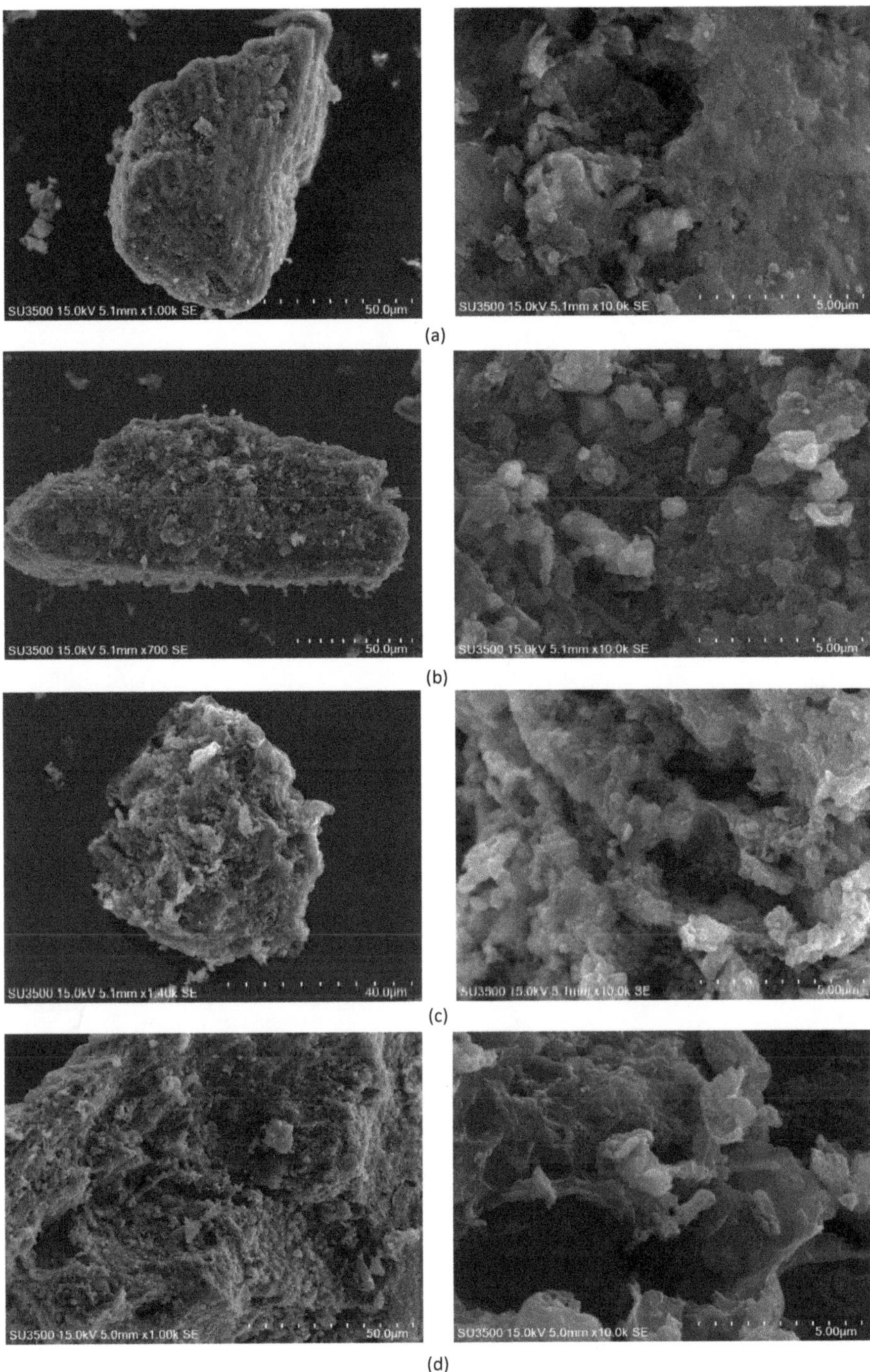

Figure 5. SEM images of lignite chars generated under carbon dioxide atmosphere at 800 °C and under the pressure of: (**a**) 0.1 MPa, (**b**) 1 MPa, (**c**) 2 MPa and (**d**) 3 MPa.

4. Conclusions

On the basis of the experimental study performed and presented within the paper the following conclusions may be drawn:

1. The elevated pressure and high temperature may be successfully applied instead of partial carbon dioxide oxidation for production of lignite-derived chars of well-developed surface area and relatively low average pore diameter.
2. Application of pressure in carbon dioxide-activation process at 800 °C results in generation of chars comparable or superior in terms of the specific surface area when compared to carbon materials received with demineralization and carbon dioxide activation of lignite, which implies that pressure may be considered as an alternative to demineralization preceding the carbon dioxide activation.
3. The combined pressure and carbon dioxide activation may be effectively applied in conversion of poor quality lignite into porous materials.

Funding: This research was funded by the Ministry of Science and Higher Education, Poland, grant number 10171019.

Conflicts of Interest: The author declares no conflict of interest.

References

1. BP Statistical Review of World Energy June 2017. Available online: https://www.bp.com/content/dam/bp-country/de_ch/PDF/bp-statistical-review-of-world-energy-2017-full-report.pdf (accessed on 19 January 2015).
2. Howaniec, N.; Smoliński, A. Biowaste utilization in the process of co-gasification with bituminous coal and lignite. *Energy* **2017**, *118*, 18–23. [CrossRef]
3. Smoliński, A.; Howaniec, N. Co-gasification of coal/sewage sludge blends to hydrogen-rich gas with the application of simulated high temperature reactor excess heat. *Int. J. Hydrogen Energy* **2016**, *41*, 8154–8158. [CrossRef]
4. Smoliński, A.; Howaniec, N.; Stańczyk, K. A comparative experimental study of biomass, lignite and hard coal steam gasification. *Renew. Energy* **2011**, *36*, 1836–1842. [CrossRef]
5. Karaman, I.; Yagmur, E.; Banford, A.; Aktas, Z. The effect of process parameters on the carbon dioxide based production of activated carbon from lignite in a rotary reactor. *Fuel Process. Technol.* **2014**, *118*, 34–41. [CrossRef]
6. Xing, B.; Huang, G.; Chen, L.; Guo, H.; Zhang, C.; Xie, W.; Chen, Z. Microwave synthesis of hierarchically porous activated carbon from lignite for high performance supercapacitors. *J. Porous Mater.* **2016**, *23*, 67–73. [CrossRef]
7. Carrott, P.J.M.; Cansado, I.P.P.; Mourão, P.A.M.; Ribeiro Carrott, M.M.L.; Louro, N.D.B.; Albiniak, A.; Broniek, E.; Jasieńko-Hałat, M. On the use of ethanol for evaluating microporosity of activated carbons prepared from Polish lignite. *Fuel Process. Technol.* **2012**, *103*, 34–38. [CrossRef]
8. Fukuyama, H.; Terai, S. Preparing and characterizing the active carbon produced by steam and carbon dioxide as a heavy oil hydrocracking catalyst support. *Catal. Today* **2008**, *130*, 382–388. [CrossRef]
9. Lü, G.; Hao, J.; Liu, L.; Ma, H.; Fang, Q.; Wu, L.; Wei, M.; Zhang, M. The adsorption of phenol by lignite activated carbon. *Chin. J. Chem. Eng.* **2011**, *19*, 380–385. [CrossRef]
10. Shrestha, S.; Son, G.; Lee, S.H.; Lee, T.G. Isotherm and thermodynamic studies of Zn (II) adsorption on lignite and coconut shell-based activated carbon fiber. *Chemosphere* **2013**, *92*, 1053–1061. [CrossRef]
11. Skodras, G.; Orfanoudaki, T.; Kakaras, E.; Sakellaropoulos, G.P. Production of special activated carbons from lignite for environmental purposes. *Fuel Process. Technol.* **2002**, *77–78*, 75–87. [CrossRef]
12. Chattopadhyaya, G.; Macdonald, D.G.; Bakhshi, N.N.; Mohammadzadeh, J.S.S.; Dalai, A.K. Preparation and characterization of chars and activated carbons from Saskatchewan lignite. *Fuel Process. Technol.* **2006**, *87*, 997–1006. [CrossRef]

13. Aivalioti, M.; Pothoulaki, D.; Papoulias, P.; Gidarakos, E. Removal of BTEX, MTBE and TAME from aquous solutions by adsorption onto raw and thermally treated lignite. *J. Hazard. Mater.* **2012**, *207–208*, 136–146. [CrossRef] [PubMed]
14. Tong, K.; Lin, A.; Ji, G.; Wang, D.; Wang, X. The effects of adsorbing organic pollutants from super heavy oil wastewater by lignite activated coke. *J. Hazard. Mater.* **2016**, *308*, 113–119. [CrossRef] [PubMed]
15. Nwaka, D.; Tahmasebi, A.; Tian, L.; Yu, T. The effects of pore structure on the behavior of water in lignite coal and activated carbon. *J. Colloid Interf. Sci.* **2016**, *477*, 138–147. [CrossRef] [PubMed]
16. Huang, H.; Wang, Y.; Cannon, F.S. Pore structure development of in-situ pyrolyzed coals for pollution prevention in iron foundries. *Fuel Process. Technol.* **2009**, *90*, 1183–1191. [CrossRef]
17. Samaras, P.; Dabou, X.; Sakellaropoulos, G.P. Thermal treatment of lignite for carbon molecular sieve production. *J. Therm. Anal.* **1998**, *52*, 717–728. [CrossRef]
18. Mahmoudi, K.; Hamdi, N.; Srasra, E. Study of adsorption of methylene blue onto activated carbon from lignite. *Surf. Eng. Appl. Elect.* **2015**, *5*, 427–433. [CrossRef]
19. Depci, T. Comparison of activated carbon and iron impregnated activated carbon derived from Gölbaşı lignite to remove cyanide from water. *Chem. Eng. J.* **2012**, *181–182*, 467–478. [CrossRef]
20. Zhu, S.; Bai, Y.; Luo, K.; Hao, C.; Bao, W.; Li, F. Impacts of CO_2 on char structure and the gasification reactivity. *J. Anal. Appl. Pyrol.* **2017**, *128*, 13–17. [CrossRef]
21. Howaniec, N.; Smoliński, A. Porous Structure Properties of Andropogon Gerardi Derived Carbon Materials. *Materials* **2018**, *11*, 876. [CrossRef]
22. Howaniec, N. The effects of pressure on coal chars porous structure development. *Fuel* **2016**, *172*, 118–123. [CrossRef]
23. Howaniec, N. Temperature induced development of porous structure of bituminous coal chars at high pressure. *J. Sustain. Min.* **2016**, *15*, 120–124. [CrossRef]
24. Benfell, K.E.; Liu, G.; Roberts, D.G.; Harris, D.J.; Lucas, J.A.; Bailey, J.G. Modelling char combustion: The influence of parent coal petrography and pyrolysis pressure on the structure and intrinsic reactivity of its char. *Proc. Combust. Inst.* **2000**, *28*, 2233–2241. [CrossRef]
25. Howaniec, N. Development of porous structure of lignite chars at high pressure and temperature. *Fuel Process. Technol.* **2016**, *154*, 163–167. [CrossRef]
26. Smoliński, A.; Howaniec, N. Analysis of Porous Structure Parameters of Biomass Chars Versus Bituminous Coal and Lignite Carbonized at High Pressure and Temperature—A Chemometric Study. *Energies* **2017**, *10*, 1457. [CrossRef]
27. Lee, C.W.; Jenkins, R.G.; Schobert, H.H. Structure and reactivity of char from elevated pressure pyrolysis of Illinois No. 6 bituminous coal. *Energy Fuel* **1992**, *6*, 40–47. [CrossRef]
28. Bryant, G.W.; Harris, D.; Tate, A.; Wall, T.F. *ACARP Project C6051 Final Report, Testing Australian Black Coals in a Pressurised Drop Tube Furnace*; ACARP: Brisbane, Australia, 1999.
29. Khan, M.R.; Jenkins, R.G. Swelling and plastic properties of coal devolatilized at elevated pressures: An examination of the influence of coal type. *Fuel* **1986**, *65*, 725–731. [CrossRef]
30. Lee, C.W.; Scaroni, A.W.; Jenkins, R.G. Effect of pressure on the behavior of a softening coal during rapid heating. *Fuel* **1991**, *70*, 957–965. [CrossRef]
31. Zdeb, J.; Howaniec, N.; Smoliński, A. Utilization of Carbon Dioxide in Coal Gasification—An Experimental Study. *Energies* **2019**, *12*, 140. [CrossRef]
32. Dai, P.; Dennis, J.S.; Scott, S.A. Using an experimentally—Determined model of the evolution of pore structure fr the gasification of chars by CO_2. *Fuel* **2016**, *17*, 29–43. [CrossRef]
33. Bielowicz, B. Petrographic characteristics of lignite gasification chars. *Int. J. Coal. Geol.* **2016**, *168*, 146–161. [CrossRef]
34. Salatino, P.; Senneca, O.; Masi, S. Gasification of coal char by oxygen and carbon dioxide. *Carbon* **1997**, *36*, 443–452. [CrossRef]
35. Grigore, M.; Sakurovs, R.; French, D.; Sahajwalla, V. Mineral reactions during coke gasification with carbon dioxide. *Int. J. Coal. Geol.* **2008**, *75*, 213–224. [CrossRef]
36. Brunauer, S.; Emmett, P.; Teller, E. Adsorption of gases in multimolecular layers. *J. Am. Chem. Soc.* **1938**, *60*, 309–319. [CrossRef]
37. Lowell, S.; Shields, J.E.; Thomas, M.A.; Thommes, M. *Characterization of Porous Solids and Powders: Surface Area, Pore Size and Density*; Kluwer Academic Publishers: Dordrecht, The Netherlands, 2004.

38. Bottani, E.J.; Tascon, J.M.D. *Adsorption by Carbons*; Elsevier Ltd.: Oxford, UK, 2008.
39. Thommes, M.; Kaneko, K.; Neimark, A.V.; Olivier, J.P.; Rodriguez-Reinoso, F.; Rouquerol, J.; Sing, K.S.W. Physisorption of gases, with special reference to the evaluation of surface area and pore size distribution (IUPAC Technical Report). *Pure. Appl. Chem.* **2015**, *87*, 1051–1069. [CrossRef]

Article

CO Adsorption Performance of CuCl/Activated Carbon by Simultaneous Reduction–Dispersion of Mixed Cu(II) Salts

Cailong Xue, Wenming Hao, Wenping Cheng, Jinghong Ma * and Ruifeng Li

College of Chemistry and Chemical Engineering, Taiyuan University of Technology, Taiyuan 030024, China; xuecailong35@163.com (C.X.); haowenming@tyut.edu.cn (W.H.); chengwenping@tyut.edu.cn (W.C.); rfli@tyut.edu.cn (R.L.)

* Correspondence: majinghong@tyut.edu.cn; Tel.: +86-351-6111353

Received: 15 April 2019; Accepted: 14 May 2019; Published: 16 May 2019

Abstract: CO is a toxic gas discharged as a byproduct in tail gases from different industrial flue gases, which needs to be taken care of urgently. In this study, a CuCl/AC adsorbent was made by a facile route of physically mixing $CuCl_2$ and $Cu(HCOO)_2$ powder with activated carbon (AC), followed by heating at 533 K under vacuum. The samples were characterized by X-ray powder diffraction (XRD), inductively coupled plasma optical emission spectrometry (ICP-OES), N_2 adsorption/desorption, and scanning electron microscopy (SEM). It was shown that Cu(II) can be completely reduced to Cu(I), and the monolayer dispersion threshold of CuCl on AC support is 4 mmol·g^{-1} AC. The adsorption isotherms of CO, CO_2, CH_4, and N_2 on CuCl/AC adsorbents were measured by the volumetric method, and the CO/CO_2, CO/CH_4, and CO/N_2 selectivities of the adsorbents were predicted using ideal adsorbed solution theory (IAST). The obtained adsorbent displayed a high CO adsorption capacity, high CO/N_2, CO/CH_4, and CO/CO_2 selectivities, excellent ad/desorption cycle performance, rapid adsorption rate, and appropriate isosteric heat of adsorption, which made it a promising adsorbent for CO separation and purification.

Keywords: CuCl/AC adsorbent; CO adsorption; monolayer dispersion; isosteric heat; adsorption isotherms

---

1. Introduction

With the rapid development of C−1 chemistry recently in the chemical industry, carbon monoxide (CO) as a significant resource has been widely applied to prepare a large variety of chemical products, such as formic acid, acetic acid, oxalic acid ester, carbonic acid two methyl ester, anhydride, etc. [1,2]. Most of these preparations need high-purity CO. The main methods of producing CO are the steam reforming of natural gas and coal gasification [3]. In addition, a significant amount of CO is discharged as a byproduct in tail gases from different industrial flue gases including carbon black tail gas, silicon carbide furnace gas, yellow phosphorus tail gas, coke oven gas, blast furnace gas, etc. [4–6]. From both processes, the obtained CO is mixed with N_2, H_2, CH_4, CO_2, and vapor. CO is toxic to humans because it combines with hemoglobin in the blood to form carboxy-hemoglobin hindering the transportation and release of oxygen in the blood, which leads to death [7]. Moreover, even a trace amount of CO can poison the noble catalysts, such as the proton-exchange membrane fuel cells, which restrict the CO content below 0.2 ppm to protect the platinum electrocatalyst [8,9]. Thus, the separation and purification of CO from different gas mixtures have significance both industrially and environmentally.

Among the proven technologies for CO separation and purification, adsorption processes, such as pressure swing adsorption (PSA) and temperature swing adsorption (TSA) have the advantages of convenient operation, low energy consumption, low operating cost, etc., and have been widely used in CO separation [10–14]. Adsorbent plays a crucial role in the adsorption based gas separation process, it has been found that many porous materials, such as activated carbons [15,16], zeolites [17,18], and metal-organic frameworks (MOFs) [19,20], have adsorption capacities to a certain extent. However, it is difficult to separate and purify CO from gas mixtures by using these materials directly since their adsorption capacity and selectivities are low. Cu(I) adsorbents for CO separation have received extensive attention for their high CO adsorption capacity and high selectivity, since CO molecules can form a π-complexation bond with Cu(I) ions on the adsorbent, which are stronger than the interaction caused by van der Waals forces [21–26]. More importantly, π-complexation bonds are still weak enough to be broken by normal engineering operations, such as increasing temperature or reducing pressure, and the adsorbed CO can be easily desorbed, which makes it a suitable adsorbent in PSA and TSA systems [23]. Two approaches are used for making Cu(I) adsorbents. In the first process, Cu(I) adsorbents are prepared by impregnating Cu(II) salts into a porous support including zeolites, activated carbons (ACs) and MOFs, etc. [20,25,27], and then reducing Cu(II) to Cu(I) using reducing gases, such as H_2 or CO. However, it is difficult to control the reduction degree, and Cu(II) is easily over reduced to Cu. In the second process, Cu(I) adsorbents are prepared by direct dispersion and impregnation of CuCl. Hirai et al. [28,29] and Tamon et al. [30] obtained Cu(I)/AC adsorbents by using dispersing reagents, such as concentrated hydrochloric acid or organic solvents, to disperse CuCl onto the AC surfaces, and then drying at 403 K in N_2. Xie et al. [31] prepared CuCl/zeolite adsorbents by dispersing CuCl powder spontaneously onto the surfaces of zeolites at 623 K in an inert atmosphere, which displayed high adsorption capacity and selectivity for CO. When using CuCl as a starting material, the adsorbent preparation has to be carefully performed in a dry inert atmosphere, to prohibit the oxidation and hydrolysis of Cu(I). In our previous work, we successfully obtained Cu(I) π-complexation adsorbents with an aqueous solution of equimolar $CuCl_2$ and $Cu(HCOO)_2$ as starting materials by the traditional impregnation method followed by activating at the temperature of 583 K [32].

Herein, the purpose of this work is to develop CO adsorbent using a solid-state auto reduction–dispersion method with $CuCl_2$ and $Cu(HCOO)_2$ as the initial material. Then, X-ray powder diffraction (XRD), inductively coupled plasma optical emission spectrometry (ICP-OES), N_2 adsorption/desorption and scanning electron microscopy (SEM) were employed to characterize the samples. Pure component CO, CO_2, N_2, and CH_4 adsorption isotherms on the adsorbents were measured in a volumetric method. The CO/CO_2, CO/CH_4, and CO/N_2 selectivities of the adsorbents were predicted by using ideal adsorbed solution theory (IAST). The adsorption isotherms were fitted with the Langmuir–Freundlich model, and the corresponding heats of adsorption were calculated. The cyclic CO adsorption on adsorbent was performed to evaluate its repeated availability during the adsorption and desorption cycles. Furthermore, the CO adsorption rate on adsorbent was discussed and reported.

2. Materials and Methods

2.1. Materials

Copper formate tetrahydrate ($Cu(HCOO)_2{\cdot}4H_2O$, 98%) and cupric chloride dihydrate ($CuCl_2{\cdot}2H_2O$, 99%) were purchased from Alfa Aesar Chemical Co. Ltd.(Ward Hill, MA, USA). Activated carbon (AC) was purchased from Chengde Jingda Activated Carbon Manufacturing Co. Ltd. (Chengde, China).

2.2. Preparation of CuCl/AC Adsorbents

CuCl/AC adsorbents were synthesized following two steps. First, the AC was physically mixed with $CuCl_2$ and $Cu(HCOO)_2$ powder to obtain CuCl/AC adsorbent precursors. Then, the obtained precursors were dried at 373 K and activated in a tube furnace at 533 K for 4 h under vacuum. The obtained precursors and CuCl/AC adsorbents were marked as Cu(II)-x/AC and Cu(I)-x/AC (x = 2, 3, 4, 5, 6), in which the loading of copper is 2, 3, 4, 5, and 6 mmol·g^{-1} AC, respectively. The as-synthesized CuCl/AC adsorbents were stored in vacuum dry storage in a desiccator.

2.3. Adsorbent Characterization

Powder X-ray diffraction (XRD) patterns of the samples were recorded by a Shimadzu LabX XRD-6000 system (Kyoto, Japan) in the 2θ range of 5 to 35° using CuKα1 (λ=1.54056 Å) radiation operated at 40 kV and 30 mA. The pore volume and surface area of the samples were calculated from N_2 adsorption/desorption isotherms measured on a surface area and pore size analyzer (QUADRASORB SI, Quantachrome Inc., Boynton Beach, FL, USA) after activating the samples at 393 K for 4 h under vacuum. The specific surface areas (S_{BET}) were determined using the BET (Brunauer–Emmett–Teller) method under relative pressure in the range of 0.01 to 0.20. The adsorbed amount of N_2 at $p/p_0 = 0.98$ was employed to calculate the total pore volume (V_{Total}). Scanning electron microscope (SEM, Hitachi S4800, Hitachi Ltd., Tokyo Japan) was used to observe the samples' morphology. Cu contents were measured by inductively coupled plasma optical emission spectrometry (ICP-OES, Thermo-ICAP6300, Thermo Fisher Scientific Co., Ltd., Waltham, MA, USA).

2.4. Adsorption Measurements

Before the adsorption measurements, the samples were degassed under vacuum while heating up to 393 K for 4 h. The CO, CO_2, CH_4, and N_2 adsorption isotherms were measured at the temperature required using a static volumetric apparatus (NOVE1000e, Quantachrome Inc., Boynton Beach, FL, USA). During the adsorption measurements, the temperature was maintained by circulating ethanediol-water from a bath with setting temperature. The adsorption capacity was determined from the adsorption isotherm measured at 298 K. Ultrahigh purity grade CO (99.99%), CO_2 (99.999%), CH_4 (99.99%), and N_2 (99.999%) were used without any purification.

3. Results and Discussion

3.1. Characterization of Samples

The XRD patterns of the copper loaded AC samples before and after activation are presented in Figure 1. Before activation, the diffraction peaks of $Cu(HCOO)_2$ and $CuCl_2$ [33,34] can be observed in the Cu(II)-x/AC samples, and the reflection intensities increased with the increase of copper loading. After activation, the diffraction peaks of $CuCl_2$ and $Cu(HCOO)_2$ disappeared, and the Cu(I)-5/AC sample displayed only a relatively weak peak (2θ = 28.5°) of CuCl [35], suggesting that $Cu(HCOO)_2$ and $CuCl_2$ were transformed into CuCl after activation. Meanwhile, the absence of CuCl diffraction peak on Cu(I)-2/AC, Cu(I)-3/AC, and Cu(I)-4/AC samples might be due to the well dispersion of CuCl on the AC surface beyond the detection limit of XRD [36]. By further increasing the copper loading to 5 and 6 mmol·g^{-1}, the appearance of the characteristic peak of CuCl implies that the crystal size of CuCl on the AC surface increased with the increase of CuCl loading, which was able to be detected by XRD with the CuCl loading higher than 5 mmol·g^{-1}.

Figure 2a,b show the representative SEM images of the copper loaded AC samples before and after activation. It can be clearly observed that the particles of copper species present on the AC surface for the Cu(II)-4/AC. However, the particles on the AC surface disappeared after activation, which implies that the activation process contributes to the CuCl dispersion on the AC surface. Figure 2c,d show selected-area and the element mapping analyses of Cu(I)-4/AC. It revealed that the copper particles are uniformly dispersed on the AC surface. The good dispersion of CuCl on Cu(I)-4/AC observed by SEM agreed well with the XRD results in Figure 1.

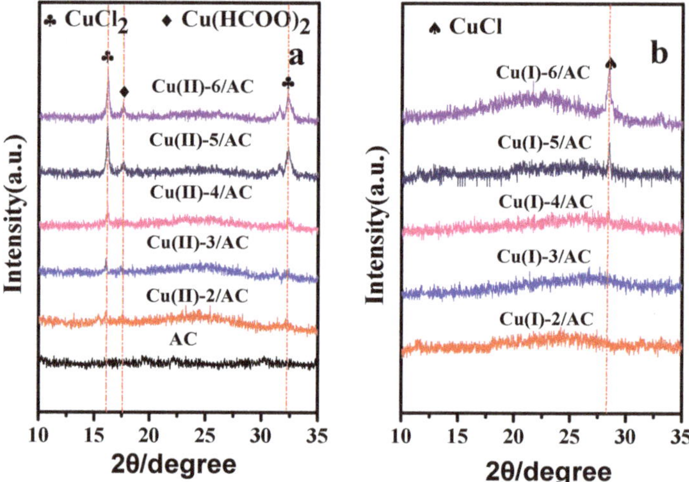

Figure 1. X-ray powder diffraction (XRD) patterns of activated carbon (AC), Cu (II)-x/AC (**a**) and Cu(I)-x/AC (**b**).

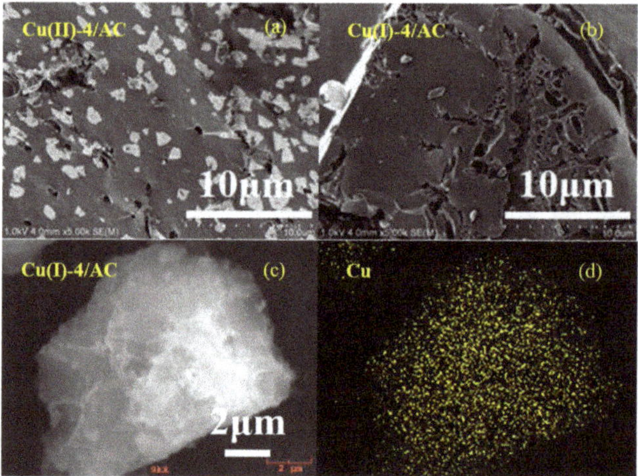

Figure 2. Scanning electron microscopy (SEM) images of Cu(II)-4/AC (**a**), Cu(I)-4/AC (**b**), and selected-area element mapping analyses of Cu(I)-4/AC (**c**,**d**).

Figure 3 shows the N_2 adsorption/desorption isotherms of the Cu(I)-x/AC samples at 77 K. The N_2 adsorption gradually decreased with increasing CuCl loading. Table 1 lists the textural parameters of AC and Cu(I)-x/AC. It can be observed that the total pore volume (V_{Total}) and BET surface area (S_{BET}) gradually decreased with the increase of CuCl loading, indicating that CuCl had been loaded into the pores of the parent AC. As the CuCl loading increased, more and more surface within the pores were occupied by CuCl, which may result in a further decrease of V_{Total} and S_{BET}. As CuCl was well dispersed on the surface of AC when the CuCl loading was below 4 mmol/g, the average pore sizes decreased with the increase of CuCl loading. Smaller pores were filled with the further increase of the CuCl loading. Therefore, the increased average pore size resulted from the percentage increase of the available larger pores in AC, which is similar to the observation by Ramli et al. [37].

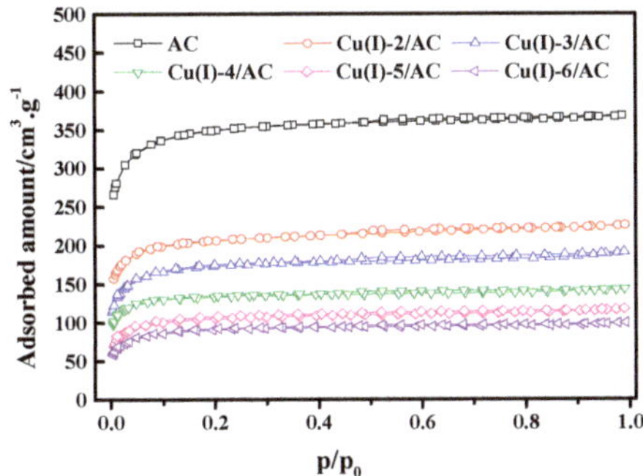

Figure 3. N_2 adsorption/desorption isotherms of Cu(I)-x/AC.

Table 1. The parameters of pore structure, loading and utilization coefficient of CuCl for Cu(I)-x/activated carbon (AC).

Samples	S_{BET} (m^2/g)	V_{Total} (cm^3/g)	d[a] (nm)	N-Cu[b] (wt %)	Cu[c] (wt %)	η (%)
AC	1082	0.571	2.11	/	/	/
Cu(I)-2/AC	804	0.372	1.85	10.7	10.2	72.5
Cu(I)-3/AC	644	0.294	1.83	14.8	14.1	69.0
Cu(I)-4/AC	505	0.223	1.76	18.3	17.5	68.9
Cu(I)-5/AC	395	0.182	1.84	21.3	20.0	58.0
Cu(I)-6/AC	331	0.156	1.89	24.0	22.5	43.6

[a] average pore diameter of adsorbent; [b] calculated from starting material; [c] obtained by ICP-OES.

It can be observed from Figure 4 that the CO adsorption capacity of CuCl/AC increased with CuCl loading in the range of 0 to 4 mmol·g^{-1}. The maximum value of adsorption capacity was 45.4 cm^3·g^{-1}. With the continuous increase of the CuCl loading, the CO adsorption capacity of CuCl/AC with the copper loading of 5 mmol·g^{-1} was almost the same as that with 4 mmol·g^{-1}. The decrease of CO adsorption occurred with further increasing the copper loading to 6 mmol·g^{-1}. This phenomenon can be ascribed to the following reason. The more copper loaded, the more active sites of CuCl/AC adsorbents present, which would enhance CO adsorption. However, the increase of copper loading also resulted in a decrease of surface area for CuCl/AC adsorbents, as shown in Table 1. As a result, the increase of the adsorbed amount from the increased active sites and the decrease of adsorbed amount from the decrease of surface area were in a dynamic balance in the copper loading range of 4 to 5 mmol·g^{-1}. When the copper loading reached 6 mmol·g^{-1}, on the one hand, the amount of adsorbed CO decreased because the decrease of surface area; on the other hand, the Cu(I) started to agglomerate on AC surface with considerable copper loading, resulting in the low utilization of active sites.

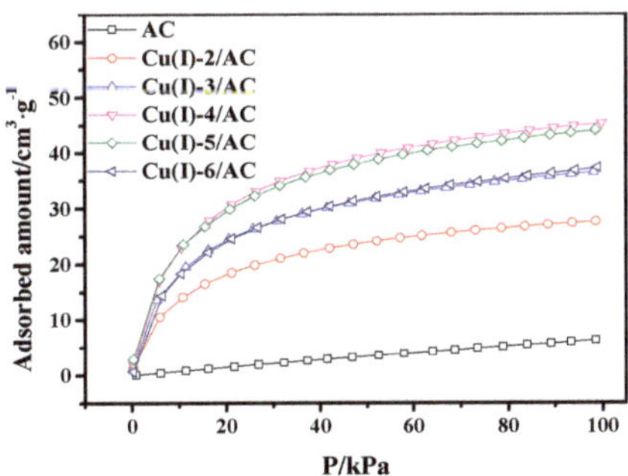

Figure 4. CO adsorption isotherms on AC and Cu(I)-x/AC at 298 K.

In addition, Table 1 lists the measured Cu contents with ICP-OES, which are closely approximate to the values in the raw material. The utilization coefficient of surface CuCl is described from the equation:

$$\eta = \frac{q_{CO}}{n_{CuCl}} \times 100\%$$

where η is the utilization coefficient of CuCl, q_{CO} is the actual CO adsorption capacity at 298 K and 100 kPa, n_{CuCl} is the mole of CuCl per gram CuCl/AC adsorbent. According to this equation, the utilization coefficients were calculated and presented in Table 1. It can be seen that the utilization coefficient decreased with the increasing of CuCl loading, which means that the high CuCl loading on AC could not guarantee high utilization of CuCl, since not all Cu(I) species can be utilized.

3.2. Adsorption Selectivities of CO to CO_2, CH_4, and N_2

Figure 5a gives the adsorption isotherms of pure CO, CO_2, CH_4, and N_2 on Cu(I)-4/AC in the pressure range of 0 to 100 kPa. The adsorption of CO_2, CH_4, and N_2 on Cu(I)-4/AC increased almost linearly with pressure, while the adsorption isotherm of CO on Cu(I)-4/AC presented a type-I isotherm [38], that is the CO adsorption increased sharply with pressure at a low pressure range, implying the adsorption of relatively strong CO-Cu(I) π-complexation, which is propitious to separate CO from $CO/CO_2/CH_4/N_2$ mixed gas.

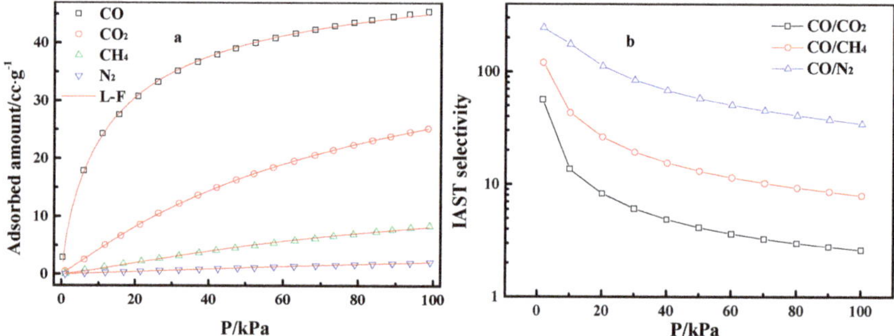

Figure 5. Adsorption isotherms of CO, CO_2, CH_4, and N_2 on Cu(I)-4/AC and Langmuir–Freundlich (L-F) fitting lines (**a**), and ideal adsorbed solution theory (IAST)-predicted adsorption selectivities (**b**).

The Langmuir–Freundlich (L-F) model and IAST were employed together to calculate the CO/CO_2, CO/CH_4, and CO/N_2 selectivities with the equimolar CO/CO_2, CO/CH_4, and CO/N_2 mixture. The L-F model can be expressed as

$$q = q_m \frac{bp^{1/n}}{1 + bp^{1/n}}$$

where q is the adsorbed amount, p is the pressure and q_m is the saturation adsorbed amount, b is the adsorption affinity and n is the corresponding deviation from the Langmuir isotherm.

First, the adsorption isotherm of pure CO, CO_2, CH_4, and N_2 were fitted by the L-F model [38]. After that, the CO/CO_2, CO/CH_4, and CO/N_2 selectivities were predicted by IAST theory [39,40]. Finally, the relevant selectivities curves along with the increase of pressure were obtained. Figure 5b shows that the CO/CO_2, CO/CH_4, and CO/N_2 selectivities decrease gradually with increasing pressure. Nevertheless, the CO/CO_2, CO/CH_4, and CO/N_2 selectivities on Cu(I)-4/AC were still up to 2.6, 8.0, and 34.3 at 100 kPa, respectively, which suggests that it has the potential for the effective separation of CO from the gas mixtures. Table 2 lists the benchmark materials for CO adsorption. Cu(I) adsorbents have higher adsorption capacity than the conventional porous adsorbent. The Cu(I)-4/AC adsorbent prepared in this study has relatively high CO/CO_2 selectivity among the selected adsorbents.

Table 2. Comparisons with adsorbents in the literature.

Adsorbent	Refs.	T[a](K)	q^b_{100} (cm^3·g^{-1})				Selectivities		
			CO	CO_2	CH_4	N_2	CO/CO_2	CO/CH_4	CO/N_2
5A	[1]	298	26.9						
13x	[1,12]	298	13.5	102	8.9	5.0	0.1	1.5	2.7
BPL AC	[14]	298	4.4	24.4	8.5		0.2	0.5	
AC	[13]	303	11.5	58	24.8	7.4	0.2	0.5	1.6
CuCl(5)/Y	[22]	303	66.9	24.1	6.7	1.0	2.8	10.0	66.9
Cu(I)/AC	[32]	298	56.0	46.2	9.6	2.5	1.2	5.8	22.4
CuCl/NaY	[31]	303	52.0	29.3	3.9	1.8	1.7	13.3	28.8
CuCl/13X	[31]	303	84.9	53.1			1.6		
Cu(I)-4/AC	This work	298	45.4	25.2	8.4	2.1	2.6	3.0	34.3

[a] the temperature of adsorption measurements; [b] adsorbed amount of gases at 100 kPa.

3.3. Isosteric Heat of Adsorption

Isosteric heat of adsorption is a significant thermodynamic parameter to characterize the interaction between the adsorbate and the adsorbent and to design a gas adsorption separation process, which can be calculated by Clausius–Clapeyron equation [41] as

$$\left[\frac{\partial lnP}{\partial(1/T)}\right]_q = -\frac{\Delta H_s}{RT}$$

where P is the pressure, R is the ideal gas constant, T is the experimental temperature, q is the adsorption amount, and ΔH_s is the isosteric heat of adsorption. In this work, the experimental isotherms and the L-F model predicted isotherms of CO at different temperatures of 273 K, 293 K, and 298 K (as shown in Figure 6) were used to calculate ΔH_s of CO adsorption on AC and Cu(I)-4/AC. The conventional Langmuir–Freundlich (L-F) adsorption model correlated the experimental results and the fitting parameters, which are listed in Table 3. The experimental data fit well with L-F model, as can be seen by the high values of R^2 (the coefficient of the experimental data and the fitting data). ΔH_s can be derived from the slopes of the plots of lnP versus $1/T$ at given adsorption amounts, as shown in Figure 7. It was shown that the isosteric heats of CO adsorption on Cu(I)-4/AC are remarkably much higher than those on AC. The result indicates that the π-complexation interaction between CO and Cu(I) is stronger than the van der Waals interaction of CO with the parent adsorbent. Usually, ΔH_s is <20 kJ·mol^{-1} for common physical adsorption and >80 kJ·mol^{-1} for chemical adsorption [42]. The values of ΔH_s on Cu(I)-4/AC maintained around 50 kJ·mol^{-1} in the whole pressure range, suggesting that the strength of complex adsorption is between physisorption and chemisorption. Such isosteric heat is not only propitious to adsorb CO, but also liable to desorb CO with a normal engineering operations (evidence as shown in Figure 9).

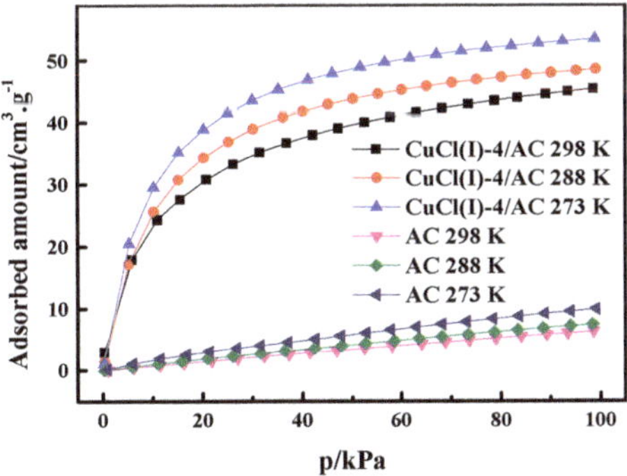

Figure 6. CO adsorption isotherms of AC and Cu(I)-4/AC at 273 K, 288 K, and 298 K.

Table 3. Langmuir–Freundlich (L-F) fitting parameters of CO isotherms on AC and Cu(I)-4/AC.

T(K)	AC				CuCl(I)-4/AC			
	q_m	b	n	R^2	q_m	b	n	R^2
298	10.22	1.82×10^{-3}	0.688	0.994	51.87	0.0945	1.095	0.999
288	15.12	3.18×10^{-3}	0.809	0.997	55.34	0.102	1.081	0.999
273	20.02	5.64×10^{-3}	0.902	0.994	60.19	0.115	1.085	0.999

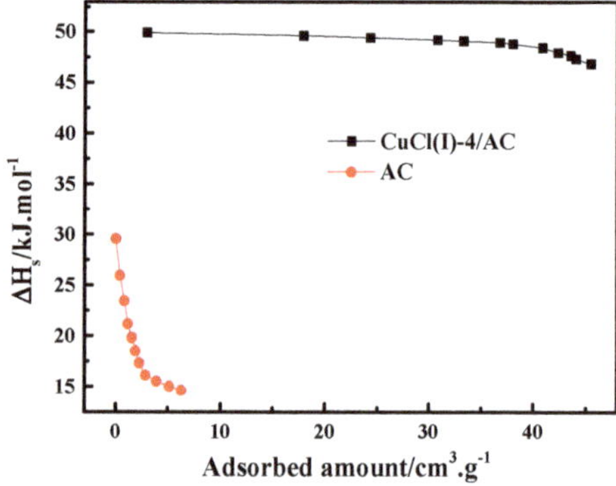

Figure 7. Isosteric heats of AC and Cu(I)-4/AC as the function of the adsorbed amount of CO.

3.4. Adsorption Kinetics of CO

In addition, it is also crucial that the CO adsorption rate need to be quite rapid for potential applications of CuCl/AC adsorbent in the adsorption-driven separation of CO from gas mixtures containing CO, CO_2, CH_4, and N_2. Typically, the requirement of the adsorption process in industrial applications is shorter than 1 min [43]. Here, we studied the time-dependent adsorption of CO on Cu(I)-4/AC adsorbent by releasing a small amount of CO and studying the adsorbed amount as a function of time as shown in Figure 8. Cu(I)-4/AC showed a relatively rapid adsorption rate, which reached 96% of the CO capacity within 25 s. The rapid CO adsorption rate suggests that Cu(I)-4/AC can meet the requirements for industrial application of adsorbent to separate CO from gas mixtures containing CO, CO_2, CH_4, and N_2 in a PSA process.

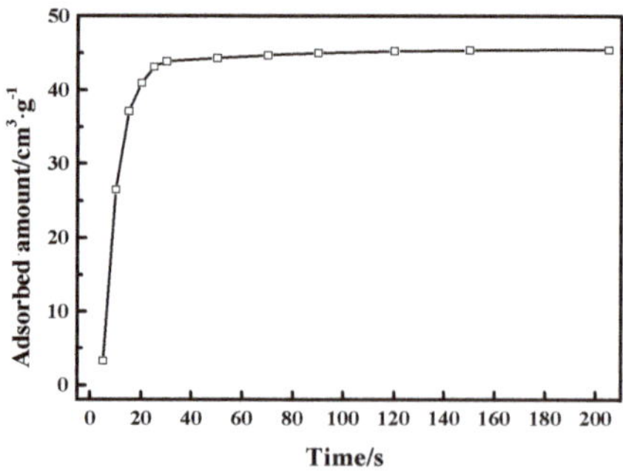

Figure 8. Uptake kinetics for Cu(I)-4/AC at 298 K.

3.5. Cycle Adsorption of CO on Cu(I)/AC

In the actual processes of gas separation, an ideal adsorbent not only needs to have high adsorption capacity and high selectivity but also needs to exhibit a stable cyclic adsorption performance in long-term adsorption/desorption cyclical operation. The pure CO cyclical adsorption/desorption isotherm at 298 K was evaluated for six times (The degassing between each cycle was carried out at 353 K under vacuum). As shown in Figure 9, the maximum amount of CO adsorption was reduced by about 1.3% after six cycles of adsorption and desorption, suggesting that the CO adsorption process using CuCl/AC adsorbent is stable under the investigated conditions. Its stable adsorption behavior indicates that the CuCl/AC has broad application prospects in selective adsorption of CO. It must be noted that the parent gases for the separation of CO must be pretreated to remove moisture in industrial processes, since the Cu(I) in the CuCl/AC can be oxidized to Cu(II) in the form of copper chloride once in contact with water vapor and O_2 (especially under light condition [44,45]), which then cannot form complexation with CO.

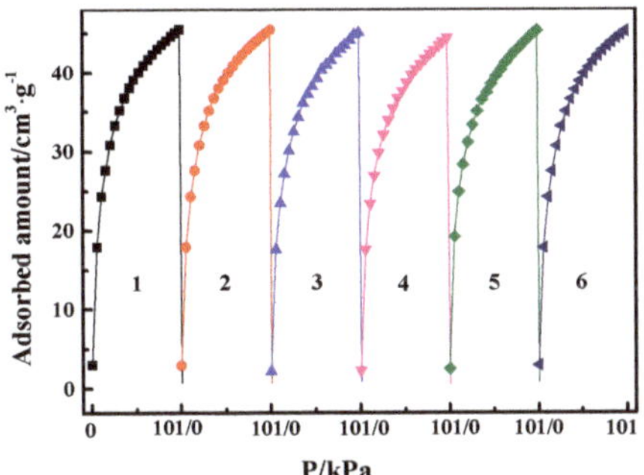

Figure 9. Adsorption and desorption cycles for CO at 298 K on Cu(I)-4/AC (the degassing between each cycle was carried out at 353 K under vacuum).

4. Conclusions

CuCl/AC adsorbents for the separation of CO have been successfully obtained using $CuCl_2$ and $Cu(HCOO)_2$ as the initial material by a solid-state auto dispersion method. $CuCl_2$ and $Cu(HCOO)_2$ can be transformed into highly dispersed CuCl with activation at 533 K under vacuum atmosphere. The CO adsorption capacity increased with transformed CuCl loading until 4 mmol·g^{-1} and then decreased afterward. The CO adsorption capacity of Cu(I)-4/AC achieved 45.4 cm^3·g^{-1}, and the CO/CO_2, CO/CH_4, and CO/N_2 selectivities were up to 2.6, 8.0, and 34.3 at 100 kPa, respectively. In addition, the isosteric heat of adsorption on Cu(I)-4/AC was about 50 kJ·mol^{-1}. The CO adsorption capacity almost remains constant during six times cyclical adsorption and rapid adsorption kinetics at the adsorption process. Those excellent properties of Cu(I)-4/AC adsorbent would make it a promising adsorbent for CO separation and purification.

Author Contributions: Conceptualization, J.M.; methodology, C.X. and J.M.; formal analysis, C.X., W.H. and W.C.; resources, R.L.; data curation, W.C.; writing—original draft preparation, C.X.; writing—review and editing, W.H.; supervision, J.M.; funding acquisition, J.M. and R.L.

Funding: This research was funded by Key Scientific and Technological Project of coal fund of Shanxi province (No.FT201402-03) and Shanxi Provincial Key Innovative Research Team in Science and Technology (No.2014131006).

Acknowledgments: This work was supported by Key Scientific and Technological Project of coal fund of Shanxi province (No.FT201402-03) and Shanxi Provincial Key Innovative Research Team in Science and Technology (No.2014131006).

Conflicts of Interest: The authors declare no conflict of interest.

References

1. Saha, D.; Deng, S. Adsorption equilibria and kinetics of carbon monoxide on zeolite 5A, 13X, MOF-5, and MOF−177. *J. Chem. Eng. Data* **2009**, *54*, 2245–2250. [CrossRef]
2. Wu, X.F.; Fang, X.; Wu, L.; Jackstell, R.; Neumann, H.; Beller, M. Transition-metal-catalyzed carbonylation reactions of olefins and alkynes: A personal account. *Acc. Chem. Res.* **2014**, *47*, 1041–1053. [CrossRef] [PubMed]
3. Heymans, N.; Alban, B.; Moreau, S.; De Weireld, G. Experimental and theoretical study of the adsorption of pure molecules and binary systems containing methane, carbon monoxide, carbon dioxide and nitrogen. Application to the syngas generation. *Chem. Eng. Sci.* **2011**, *66*, 3850–3858. [CrossRef]
4. Harlacher, T.; Melin, T.; Wessling, M. Techno-economic analysis of membrane-based argon recovery in a silicon carbide process. *Ind. Eng. Chem. Res.* **2013**, *52*, 10460–10466. [CrossRef]
5. Zarca, G.; Ortiz, I.; Urtiaga, A. Kinetics of the carbon monoxide reactive uptake by an imidazolium chlorocuprate (I) ionic liquid. *Chem. Eng. J.* **2014**, *252*, 298–304. [CrossRef]
6. Gao, F.; Wang, Y.; Wang, X.; Wang, S. Selective CO adsorbent CuCl/AC prepared using $CuCl_2$ as a precursor by a facile method. *RSC Adv.* **2016**, *6*, 34439–34446. [CrossRef]
7. DeCoste, J.B.; Peterson, G.W. Metal–organic frameworks for air purification of toxic chemicals. *Chem. Rev.* **2014**, *114*, 5695–5727. [CrossRef]
8. Pérez, L.C.; Koski, P.; Ihonen, J.; Sousa, J.M.; Mendes, A. Effect of fuel utilization on the carbon monoxide poisoning dynamics of Polymer Electrolyte Membrane Fuel Cells. *J. Power Sources* **2014**, *258*, 122–128.
9. Romero, E.L.; Wilhite, B.A. Composite catalytic-permselective membranes: Modeling analysis for H_2 purification assisted by water–gas-shift reaction. *Chem. Eng. J.* **2012**, *207*, 552–563. [CrossRef]
10. Wang, L.; Zhao, J.; Wang, L.; Yan, T.; Sun, Y.Y.; Zhang, S.B. Titanium-decorated graphene oxide for carbon monoxide capture and separation. *Phys. Chem. Chem. Phys.* **2011**, *13*, 21126–21131. [CrossRef]
11. Sethia, G.; Patel, H.A.; Pawar, R.R.; Bajaj, H.C. Porous synthetic hectorites for selective adsorption of carbon dioxide over nitrogen, methane, carbon monoxide and oxygen. *Appl. Clay Sci.* **2014**, *91*, 63–69. [CrossRef]
12. Sethia, G.; Somani, R.S.; Bajaj, H.C. Adsorption of carbon monoxide, methane and nitrogen on alkaline earth metal ion exchanged zeolite-X: Structure, cation position and adsorption relationship. *RSC Adv.* **2015**, *5*, 12773–12781. [CrossRef]
13. Lopes, F.V.S.; Grande, C.A.; Ribeiro, A.M.; Loureiro, J.M.; Evaggelos, O.; Nikolakis, V.; Rodrigues, A.E. Adsorption of H_2, CO_2, CH_4, CO, N_2 and H_2O in activated carbon and zeolite for hydrogen production. *Sep. Sci. Technol.* **2009**, *44*, 1045–1073. [CrossRef]
14. Delgado, J.A.; Águeda, V.I.; Uguina, M.A.; Sotelo, J.L.; Brea, P.; Grande, C.A. Adsorption and diffusion of H_2, CO, CH_4, and CO_2 in BPL activated carbon and 13X zeolite: Evaluation of performance in pressure swing adsorption hydrogen purification by simulation. *Ind. Eng. Chem. Res.* **2014**, *53*, 15414–15426. [CrossRef]
15. Bastos-Neto, M.; Moeller, A.; Staudt, R.; Böhm, J.; Gläser, R. Dynamic bed measurements of CO adsorption on microporous adsorbents at high pressures for hydrogen purification processes. *Sep. Purif. Technol.* **2011**, *77*, 251–260. [CrossRef]
16. Grande, C.A.; Lopes, F.V.S.; Ribeiro, A.M.; Loureiro, J.M.; Rodrigues, A.E. Adsorption of off-gases from steam methane reforming (H_2, CO_2, CH_4, CO and N_2) on activated carbon. *Sep. Sci. Technol.* **2008**, *43*, 1338–1364. [CrossRef]
17. Tsutaya, H.; Izumi, J. Carbon monoxide adsorption by zeolite. *Zeolites* **1991**, *11*, 90. [CrossRef]
18. Chakarova, K.; Hadjiivanov, K. H-bonding of zeolite hydroxyls with weak bases: FTIR study of CO and N_2 adsorption on HD-ZSM-5. *J. Phys. Chem. C* **2011**, *115*, 4806–4817. [CrossRef]
19. Chowdhury, P.; Mekala, S.; Dreisbach, F.; Gumma, S. Adsorption of CO, CO_2 and CH_4 on Cu-BTC and MIL−101 metal organic frameworks: Effect of open metal sites and adsorbate polarity. *Micropor. Mesopor. Mater.* **2012**, *152*, 246–252. [CrossRef]
20. Mishra, P.; Mekala, S.; Dreisbach, F.; Mandal, B.; Gumma, S. Adsorption of CO_2, CO, CH_4 and N_2 on a zinc based metal organic framework. *Sep. Purif. Technol.* **2012**, *94*, 124–130. [CrossRef]
21. Peng, J.; Xian, S.; Xiao, J.; Huang, Y.; Xia, Q.; Wang, H.; Li, Z. A supported Cu(I)@MIL−100 (Fe) adsorbent with high CO adsorption capacity and CO/N_2 selectivity. *Chem. Eng. J.* **2015**, *270*, 282–289. [CrossRef]
22. Gao, F.; Wang, Y.; Wang, S. Selective adsorption of CO on CuCl/Y adsorbent prepared using $CuCl_2$ as precursor: Equilibrium and thermodynamics. *Chem. Eng. J.* **2016**, *290*, 418–427. [CrossRef]

23. Khan, N.A.; Jhung, S.H. Adsorptive removal and separation of chemicals with metal-organic frameworks: Contribution of π-complexation. *J. Hazard. Mater.* **2017**, *325*, 198–213. [CrossRef]
24. Yoon, J.W.; Yoon, T.U.; Kim, E.J.; Kim, A.R.; Jung, T.S.; Han, S.S.; Bae, Y.S. Highly selective adsorption of CO over CO_2 in a Cu(I)-chelated porous organic polymer. *J. Hazard. Mater.* **2018**, *341*, 321–327. [CrossRef]
25. Jiang, W.J.; Yin, Y.; Liu, X.Q.; Yin, X.Q.; Shi, Y.Q.; Sun, L.B. Fabrication of supported cuprous sites at low temperatures: An efficient, controllable strategy using vapor-induced reduction. *J. Am. Chem. Soc.* **2013**, *135*, 8137–8140. [CrossRef]
26. Cho, K.; Kim, J.; Beum, H.T.; Jung, T.; Han, S.S. Synthesis of CuCl/Boehmite adsorbents that exhibit high CO selectivity in CO/CO_2 separation. *J. Hazard. Mater.* **2018**, *344*, 857–864. [CrossRef]
27. Wang, Y.; Yang, R.T.; Heinzel, J.M. Desulfurization of jet fuel by π-complexation adsorption with metal halides supported on MCM-41 and SBA−15 mesoporous materials. *Chem. Eng. Sci.* **2008**, *63*, 356–365. [CrossRef]
28. Hirai, H.; Wada, K.; Komiyama, M. Active carbon-supported copper (I) chloride as solid adsorbent for carbon monoxide. *Bull. Chem. Soc. Jpn.* **1986**, *59*, 2217–2223. [CrossRef]
29. Hirai, H.; Wada, K.; Kurima, K.; Komiyama, M. Carbon monoxide adsorbent composed of copper (I) chloride and polystyrene resin having amino groups. *Bull. Chem. Soc. Jpn.* **1986**, *59*, 2553–2558. [CrossRef]
30. Tamon, H.; Kitamura, K.; Okazaki, M. Adsorption of carbon monoxide on activated carbon impregnated with metal halide. *AIChE J.* **1996**, *42*, 422–430. [CrossRef]
31. Xie, Y.; Zhang, J.; Qiu, J.; Tong, X.; Fu, J.; Yang, G.; Yan, H.; Tang, Y. Zeolites modified by CuCl for separating CO from gas mixtures containing CO_2. *Adsorption* **1997**, *3*, 27–32. [CrossRef]
32. Ma, J.; Li, L.; Ren, J.; Li, R. CO adsorption on activated carbon-supported Cu-based adsorbent prepared by a facile route. *Sep. Purif. Technol.* **2010**, *76*, 89–93. [CrossRef]
33. Bastidas, D.M.; La Iglesia, V.M.; Cano, E.; Fajardo, S.; Bastidas, J.M. Kinetic study of formate compounds developed on copper in the presence of formic acid vapor. *J. Electrochem. Soc.* **2008**, *155*, C578–C582. [CrossRef]
34. *Powder Diffraction File (PDF) Database*; International Centre for Diffraction Data: Swarthmore, PA, USA, 1988.
35. Zhong, L.; Ruiyu, W.; Huayan, Z.; Kechang, X. Preparation of CuIY catalyst using $CuCl_2$ as precursor for vapor phase oxidative carbonylation of methanol to dimethyl carbonate. *Fuel* **2010**, *89*, 1339–1343. [CrossRef]
36. Chen, Y.; Xie, C.; Li, Y.; Song, C.; Bolin, T.B. Sulfur poisoning mechanism of steam reforming catalysts: An X-ray absorption near edge structure (XANES) spectroscopic study. *Phys. Chem. Chem. Phys.* **2010**, *12*, 5707–5711. [CrossRef]
37. Ramli, N.A.S.; Amin, N.A.S. Fe/HY zeolite as an effective catalyst for levulinic acid production from glucose: Characterization and catalytic performance. *Appl. Catal. B* **2015**, *163*, 487–498. [CrossRef]
38. Huang, W.; Zhou, X.; Xia, Q.; Peng, J.; Wang, H.; Li, Z. Preparation and adsorption performance of GrO@Cu-BTC for separation of CO_2/CH_4. *Ind. Eng. Chem. Res.* **2014**, *53*, 11176–11184. [CrossRef]
39. Magnowski, N.B.K.; Avila, A.M.; Lin, C.C.H.; Shi, M.; Kuznicki, S.M. Extraction of ethane from natural gas by adsorption on modified ETS−10. *Chem. Eng. Sci.* **2011**, *66*, 1697–1701. [CrossRef]
40. Myers, A.L.; Prausnitz, J.M. Thermodynamics of mixed-gas adsorption. *AIChE J.* **1965**, *11*, 121–127. [CrossRef]
41. Hill, T.L. Statistical mechanics of adsorption. V. Thermodynamics and heat of adsorption. *J. Chem. Phys.* **1949**, *17*, 520–535. [CrossRef]
42. Gu, B.; Schmitt, J.; Chen, Z.; Liang, L.; McCarthy, J.F. Adsorption and desorption of natural organic matter on iron oxide: Mechanisms and models. *Environ. Sci. Technol.* **1994**, *28*, 38–46. [CrossRef]
43. Hao, W.; Björkman, E.; Lilliestråle, M.; Hedin, N. Activated carbons prepared from hydrothermally carbonized waste biomass used as adsorbents for CO_2. *Appl. Energy* **2013**, *112*, 526–532. [CrossRef]
44. Carlsson, B.; Wettermark, G. Optical properties of metallic copper in relation to the photochromic system CuCl(s) H_2O(l). *J. Photochem.* **1976**, *5*, 321–328. [CrossRef]
45. Carlsson, B.; Wettermark, G. The photochromic properties of the system CuCl(s)H_2O(l) in relation to the composition of the aqueous solution. *J. Photochem.* **1979**, *11*, 403–412. [CrossRef]

Article

Effect of High Pressure on the Reducibility and Dispersion of the Active Phase of Fischer–Tropsch Catalysts

Simón Yunes [1], Miguel Ángel Vicente [2], Sophia A. Korili [3] and Antonio Gil [3,*]

1 Micromeritics Instrument Corporation, 4356 Communications Drive, Norcross, GA 30093, USA; Simon.Yunes@micromeritics.com
2 GIR QUESCAT-Departamento de Química Inorgánica, Universidad de Salamanca, 37008 Salamanca, Spain; mavicente@usal.es
3 INAMAT-Departamento de Ciencias, Universidad Pública de Navarra, 31006 Pamplona, Spain; sofia.korili@unavarra.es
* Correspondence: andoni@unavarra.es; Tel.: +34-948169602

Received: 28 May 2019; Accepted: 10 June 2019; Published: 13 June 2019

Abstract: The effect of high pressure on the reducibility and dispersion of oxides of Co and Fe supported on γ-Al_2O_3, SiO_2, and TiO_2 has been studied. The catalysts, having a nominal metal content of 10 wt.%, were prepared by incipient wetness impregnation of previously calcined supports. After drying at 60 °C for 6 h and calcination at 500 °C for 4 h, the catalysts were reduced by hydrogen at two pressures, 1 and 25 bar. The metal reduction was studied by temperature-programmed reduction up to 750 °C at the two pressures, and the metal dispersion was measured by CO chemisorption at 25 °C, obtaining values between 1% and 8%. The physicochemical characterization of these materials was completed by means of chemical analysis, X-ray diffraction, N_2 adsorption-desorption at −196 °C and scanning electron microscopy. The high pressure lowered the reduction temperature of the metal oxides, improving their reducibility and dispersion. The metal reducibility increased from 42%, in the case of Fe/Al_2O_3 (1 bar), to 100%, in the case of Fe/TiO_2 (25 bar).

Keywords: Fischer–Tropsch; supported iron oxide; supported cobalt oxide; reducibility; dispersion

---

1. Introduction

Supported cobalt and iron catalysts have been extensively studied in the Fischer–Tropsch reaction for the conversion of synthesis gas obtained from natural gas due to their high activity, high selectivity to long-chain paraffins, and low activity in the formation of water [1]. Despite this huge amount of research work, there is still no agreement in the scientific community about the active phases in the reaction between CO and H_2. Thus, for example, in the case of Fe catalysts there is a high degree of consensus about the fact that its carbides, and not metallic Fe, are the active phase. Likewise, the two factors that can control the activity of the catalysts are the degree of reduction of the metallic precursor as well as the shape and size of the metal particles formed, characteristics that are related through the dispersion, the distribution of the particles on the support [2].

In general, the type and structure of the support affect the dispersion, particle size, and reducibility, and as a consequence the activity of the supported metal catalysts [2,3]. The acidity of the supports, as well as the presence of dopants, are other factors that affect the reducibility and dispersion of the metallic phase [2,4]. Other influencing preparation variables are the metal precursor, the solvent, the metal content, the method of preparation, and the pretreatments before the catalytic tests. For example, in the case of cobalt, CoO can react with the supports both during the synthesis and during the reduction treatment, resulting in various mixed compounds like $CoAl_2O_4$, Co_2SiO_4

or $CoTiO_3$ [5–10]. These mixed compounds require far too high reduction temperatures to reduce the metal. Under this way, cobalt metal can be lost by sublimation and increase the particle sizes by sintering, resulting in catalysts with lower performance. The reduction temperatures can be decreased by increasing the pressure, also allowing less cobalt to be lost, and decreasing sintering. Other possible action is the promotion with noble metals (Pd, Pt, Re, and Ru have been used) to enhance the reducibility of the oxides, to improve the metal dispersion, to reduce the catalyst deactivation, etc. [11].

The reaction conditions, such as temperatures up to 350 °C, pressures up to 55 bar, and the presence of fluids under supercritical conditions, also control the Fischer–Tropsch synthesis process [12]. It is highly recommended to characterize the catalysts under these conditions to have a more realistic view of the effect of the different variables on the properties of the solids.

The aim of this work is to evaluate the effect of pressure on the reducibility and dispersion of Co and Fe catalysts prepared from various supports. Some examples of this effect have been published previously but only for cobalt catalysts supported on carbon nanofibers [13,14]. In this work, a comparative study also considering Fe as active phase and other catalytic supports is presented.

2. Materials and Methods

2.1. Preparation of the Catalysts

The Co and Fe catalysts, with a nominal metal content of 10 wt.%, were prepared by incipient wetness impregnation of the supports. The salts used were $Co(NO_3)_2$•$6H_2O$ (Panreac, Castellar del Vallés, Barcelona, Spain) and $Fe(NO_3)_3$•$9H_2O$ (Riedel-de Haën-Honeywell, Madrid, Spain), respectively. The commercial supports used were γ-Al_2O_3 (Spheralite 505, Procatalyse, Rueil Malmaison, France), SiO_2 (Aerolyst 350, Degussa, Frankfurt, Germany), and TiO_2 (Aeroxide TiO_2 P25, Degussa). Prior to their use, all of them were calcined in air at 500 °C for 4h. After impregnation, all the catalysts were dried at 60 °C for 6 h and calcined again at 500 °C for 4 h.

2.2. Characterization Techniques

The metal content was determined by inductively coupled plasma-atomic emission spectroscopy (ICP-AES) using Varian Vista-MPX equipment with radial vision (Varian, Palo Alto, CA, USA). The crystalline structure of the catalysts was analyzed by X-ray diffraction (XRD) in a Siemens D5000 diffractometer (Siemens, Munich, Germany). The scanning electron microscopy analyses were carried out at CLPU (Salamanca, Spain) using a Carl Zeiss SEM EVO HD25 (Zeiss Microscopy, Jena, Germany). The textural characterization of the supports and the catalysts synthesized was carried out by adsorption-desorption of N_2 (Air Liquide, 99.999%) at −196 °C using a static volumetric method in Micromeritics model ASAP 2010 equipment (Norcross, GA, USA).

The temperature-programmed reduction (TPR) and the pulse chemisorption experiments were carried out in an automated and controlled Effi Microactivity-reference PID Eng & Tech, denoted as Micro Catalyst Characterization and Testing Center, MCCTC, Madrid, Spain (see Figure 1). This equipment is mainly used to carry out catalytic reactions of any kind at a pressure between atmospheric and 200 bar. The configuration of the system allows to characterize the catalyst in situ, especially when carrying out catalytic tests in which the deactivation of the catalyst takes place. Under these conditions, it is necessary to characterize the catalyst without removing it from the reactor, avoiding its contact with the atmosphere.

The equipment consists of a hot box where all the pipes, valves, furnace, and reactor are arranged, normally kept warm at a temperature up to 200 °C to prevent the condensation of vapors. The equipment is connected in line to an MKS Instruments mass spectrometer, CirrusTM single-quadrupole model, which constitutes the system's gas analysis system (Andover, MA, USA).

The system has a tubular steel reactor of 32 mm in length and 9 mm in internal diameter, with the catalytic bed placed inside it on a porous plate. The catalytic bed is fixed between two layers of quartz wool. The reactor is located inside a longitudinal opening cylindrical furnace that allows the thermal regulation of the system at a maximum working temperature of 1100 °C. A type K thermocouple placed in contact with the catalytic bed, which is part of the temperature measurement and control system, is placed inside the reactor. The flow of gases enters the reactor in a descending way and the reaction products exit from the bottom, being then introduced into the analysis system. The flow rate of all the gases entering the reaction system is measured through mass flow controllers. The gases are preheated in the hot box of the system and directed towards a six-way valve that pneumatically directs the path of the current directly to the reactor tube or to the gas outlet. In this valve, a loop (0.521 cm^3 NTP) was adapted, which allows to dose an active gas such as CO in order to determine the dispersion of the active phase of any type of catalyst.

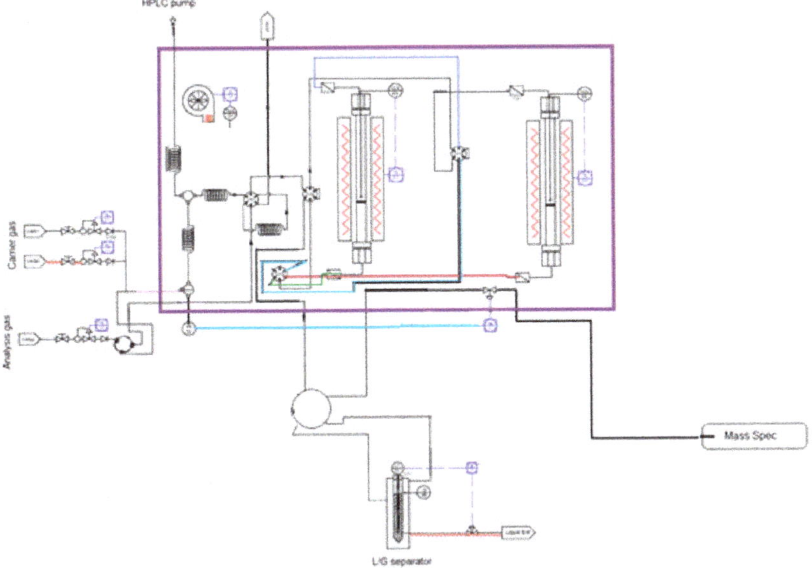

Figure 1. Flow-diagram of the Micro Catalyst Characterization and Testing Center (MCCTC) connected to a mass spectrometer.

For the TPR experiments, 0.3 g of catalyst was pretreated by heating from room temperature to 150 °C at a heating rate of 10 °C/min, and maintained at 150 °C for 2 h, all the process under a flow of N_2 (99.999%) of 50 cm^3/min. The samples were then reduced using a 2.5% H_2/Ar (99.999%) mixture with a flow of 400 cm^3/min, from room temperature to 750 °C, with a heating rate of 10 °C/min, and a pressure of 1 and 25 bar. This temperature was maintained until the baseline returned to zero, indicating a complete reduction. Two commercial pure oxides, Co_3O_4 (Sigma-Aldrich España, Madrid, Spain, 99%) and Fe_2O_3 (Sigma-Aldrich, 99.99%), were used as references for calibrating the reduction profiles and quantifying the amount of H_2 consumed by the catalysts studied.

The dispersion of the active phase was determined by repeating the reduction treatment, both at 1 and 25 bar pressure, but finishing it at 500 °C and maintaining this temperature for 4 h to ensure that the baseline was recovered. Next, the reducing mixture was replaced by an inert gas at the same temperature to ensure the elimination of all traces of H_2. In the next step, the temperature was lowered to room temperature, and the dosing was carried out using a calibrated loop of 0.521 cm^3 of CO (99.999%) until the saturation of the signal was observed, almost three identical peaks. With the volume of chemisorbed CO, the metallic content and the stoichiometric ratio CO/metal, 1:1 in this case, the percent dispersion of the metals was determined using the following Equation (1),

$$D(\%) = \frac{V_{ads} \cdot f_a \cdot W_a}{V_{mol} \cdot M(\%)} \tag{1}$$

where $D(\%)$ is the metal dispersion, V_{ads} is the total volume of CO chemisorbed, f_a is the stoichiometry factor, W_a is the atomic mass of the active metal, V_{mol} is the molar volume of the CO and $M(\%)$ is the mass percent of metal present in the catalysts.

In the case of the chemisorption temperature, the value was selected to avoid the contribution to the physical adsorption of CO on the supports. For the stoichiometric relationship, it was assumed that each molecule of CO interacted with one Co or Fe atom on the surface [15,16].

3. Results and Discussion

The XRD results indicated (see Figure 2) the formation of the cobalt spinel Co_3O_4 in the case of the cobalt catalysts and the hematite phase Fe_2O_3 in the case of iron catalysts. No mixed phases involving the supports and the active phases were detected, although in the case of alumina, Al(III) cations from the support should isomorphically incorporate to both active phases, and in fact reducibility studies strongly suggested this possibility (vide infra).

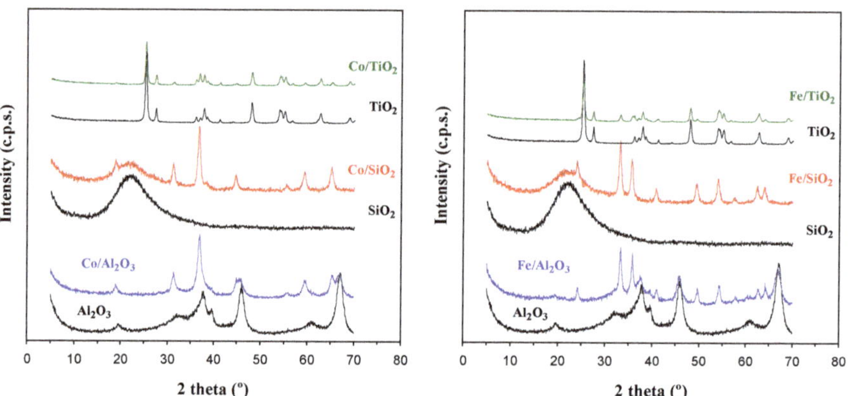

Figure 2. XRD patterns of the supported cobalt (**left**) and iron (**right**) oxide catalysts.

Selected micrographs of the cobalt catalysts are included in Figure 3. The images confirmed that the structures of the particles of the supports were not significantly modified by the impregnation with the metallic solutions and the subsequent calcination.

Figure 3. SEM images of the supported cobalt (**left**) and iron (**right**) oxide catalysts. Top: Alumina-supported catalysts; middle: Silica-supported catalysts; bottom: Titania-supported catalysts.

The adsorption isotherms were of type II according to the IUPAC classification [17] for γ-Al_2O_3 and TiO_2 supports and type IV in the case of SiO_2 support. The presence of metallic oxides did not change the shape of the adsorption isotherms or the shape of the hysteresis cycles that originated in the desorption process. This result may indicate that the metal oxides were well dispersed on the surface of the supports. The textural properties, specific surface area, and pore volume of the supports and of the catalysts are included in Table 1, in which the metal content of the catalysts is also included.

Table 1. Textural properties derived from N_2 adsorption at −196 °C and metal content by inductively coupled plasma-atomic emission spectroscopy ICP AES.

Catalyst	S_{BET} (m^2/g)	V_p (cm^3/g) *	Metal Content (wt.%)
γ-Al_2O_3	183	0.410	–
SiO_2	288	0.797	–
TiO_2	51	0.147	–
Co/Al_2O_3	122	0.291	13.54
Co/SiO_2	225	0.615	13.79
Co/TiO_2	32	0.251	10.90
Fe/Al_2O_3	132	0.290	9.11
Fe/SiO_2	220	0.562	13.30
Fe/TiO_2	38	0.235	9.62

* Total pore volume, calculated from N_2 adsorption at $p/p^o = 0.98$.

The TPR profiles corresponding to the series of Co catalysts are included in Figure 4. In all the cases, a reduction peak can be observed at 350–465 °C, which coincided with the maximum peak of reduction of pure oxide Co_3O_4 to Co^0. The reduction shoulder observed at lower temperature can be related to the reduction of Co_3O_4 to metallic Co in two steps ($Co^{3+} \rightarrow Co^{2+} \rightarrow Co^0$), as has been discussed by Arnoldy and Moulijn [18]. These results confirmed the presence of Co_3O_4. In the case of the catalyst with Al_2O_3 as support, the higher reduction temperatures may be due to a greater oxide-support interaction. Al(III) can isomorphically incorporate to the spinel phase. The peak of reduction centered at higher temperature (600–1000 °C), which was observed in the Co/Al_2O_3 catalyst, suggested the reduction of Co^{2+} species present in the form of cobalt spinels.

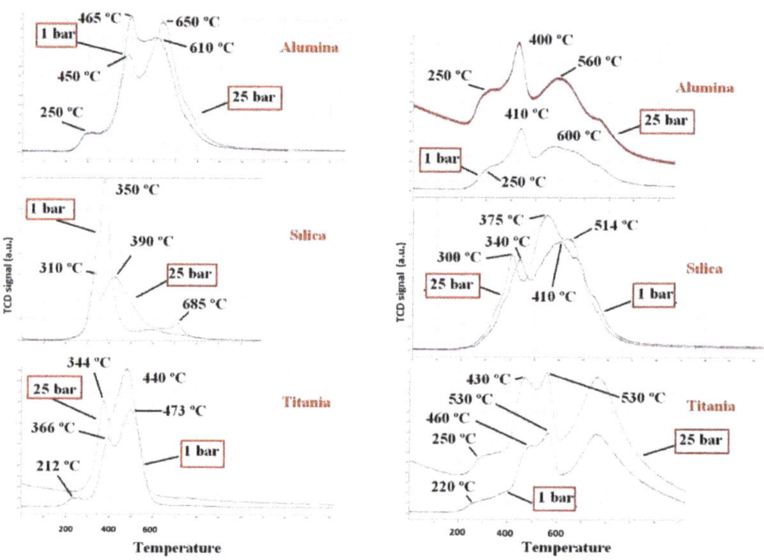

Figure 4. Reduction profiles corresponding to the supported cobalt (**left**) and iron (**right**) oxide catalysts.

In two previous works, Jacobs et al. [19] and Borg et al. [20] reported that the metal-support interactions affected the reduction of cobalt species and the strength of such interactions for various supports decreased in the order $Al_2O_3 > TiO_2 > SiO_2$. In the case of Al_2O_3 and as a result of the metal-support interactions that arose from the diffusion of cobalt ions into alumina lattice sites of octahedral or tetrahedral geometry, the formation of $CoAl_2O_4$ was proposed, and as a consequence the reducibility of the cobalt was hindered. TiO_2 is known to exhibit the strong metal-support interaction (SMSI) effect, where during the reduction of the cobalt oxide, the partial reduction of the support also takes place, encapsulating or decorating the cobalt particles [21]. The nature of hydroxyl groups and their concentration and distribution on the silica surface play the main role in the dispersion of cobalt particles. In this type of support, the formation of Co_2SiO_4 has also been reported.

The TPR profiles corresponding to the series of Fe catalysts are also included in Figure 4. The reduction of Fe_2O_3 was related to a first peak at 400–460 °C ($3Fe_2O_3 \rightarrow 2Fe_3O_4$) and a second broader peak located at high temperatures that reflected the reduction to metallic iron ($2Fe_3O_4 \rightarrow 6Fe$). The shoulder at 630 °C indicated an intermediate reduction state ($2Fe_3O_4 \rightarrow 6FeO \rightarrow 6Fe$).

The volumes of H_2 consumed by weight of metal oxide, the degree of reducibility of the catalysts, and the dispersion of the metal phases for the catalysts studied are included in Table 2. The results are given for the two working pressures, 1 and 25 bar.

Table 2. Results of H_2 consumption, metal reducibility and dispersion.

Catalyst	$Q(H_2)$ (cm^3/g_{oxide})	R (%)	D (%)
Co/TiO_2 (25 bar)	372	100	5
Co/TiO_2 (1 bar)	315	85	1
Co/SiO_2 (25 bar)	299	80	2
Co/SiO_2 (1 bar)	296	80	1
Co/Al_2O_3 (25 bar)	296	80	6
Co/Al_2O_3 (1 bar)	271	73	3
Fe/TiO_2 (25 bar)	421	100	7
Fe/TiO_2 (1 bar)	306	73	1
Fe/SiO_2 (25 bar)	346	82	3
Fe/SiO_2 (1 bar)	295	70	1
Fe/Al_2O_3 (25 bar)	236	56	8
Fe/Al_2O_3 (1 bar)	175	42	3

The low values of metal dispersion can be related to the high metal contents of the synthesized catalysts. It was also indicative of the large size of the metal particles. For the two metals considered, the catalysts synthesized using TiO_2 as support had greater degrees of oxide reduction. This result can be attributed to the weak interaction existing between the metallic oxide and the support and characteristic of the SMSI effect. The situation was very different for the catalysts supported on γ-Al_2O_3, which showed lower degrees of oxide reduction, behavior that can be explained through strong oxide-support interactions, even with the formation of spinels involving cations from the support, that is, Fe(II) or Co(II) as divalent cations, and Fe(III) or Co(III) from the precursors and Al(III) from the support as trivalent cations, this effect seeming to be more significant in the case of Co-samples [5,18]. These effects decreased with pressure, as the reducibility of the metal oxides and the metal dispersion increased.

4. Conclusions

The effects of the high pressure of on the reducibility of catalysts based on cobalt and iron oxides on three commercial supports, γ-Al_2O_3, SiO_2, and TiO_2, have been presented. Two effects have been observed, the high pressure lowered the reduction temperature, decreasing the sintering of the metal oxide particles, while the pressure improved the reducibility of the metal oxides to an almost total reduction value. These two effects gave rise to a greater dispersion of the active metal phase, which may result in an increase in the activity of the catalysts.

Author Contributions: All the authors conceived, designed, and performed the experiments, analyzed the data, and drafted the manuscript.

Funding: This research was jointly funded by the Spanish Ministry of Economy and Competitiveness (AEI/MINECO) and the European Regional Development Fund (ERDF), grant number MAT2016-78863-C2-R. A.G. thanks Santander Bank for funding through the Research Intensification program.

Conflicts of Interest: The authors declare no conflict of interest.

References

1. Davis, B.H. Fischer-Tropsch synthesis: Comparison of performances of iron and cobalt catalysts. *Ind. Eng. Chem. Res.* **2007**, *46*, 8938–8945. [CrossRef]
2. Iglesia, E. Design, synthesis, and use of cobalt-based Fischer-Tropsch synthesis catalysts. *Appl. Catal. A* **1997**, *161*, 59–78. [CrossRef]
3. Jablonski, J.M.; Okal, J.; Potoczna-Petru, D.; Krajcyk, L. High temperature reduction with hydrogen, phase composition, and activity of cobalt/silica catalysts. *J. Catal.* **2003**, *220*, 146–160. [CrossRef]
4. Zhang, J.; Chen, J.; Ren, J.; Sun, Y. Chemical treatment of γ-Al_2O_3 and its influence on the properties of Co-based catalysts for Fischer–Tropsch synthesis. *Appl. Catal. A* **2003**, *243*, 121–133. [CrossRef]

5. Chin, R.L.; Hercules, D.M. Surface spectroscopic characterization of cobalt-alumina catalysts. *J. Phys. Chem.* **1982**, *86*, 360–367. [CrossRef]
6. Van Berge, P.J.; Van e Loosdrecht, J.; Barradas, S.; Van der Kraan, A.M. Oxidation of cobalt based Fischer–Tropsch catalysts as a deactivation mechanism. *Catal. Today* **2000**, *58*, 321–334. [CrossRef]
7. Li, J.L.; Jacobs, G.; Das, T.; Zhang, Y.Q.; Davis, B. Fischer–Tropsch synthesis: Effect of water on the catalytic properties of a Co/SiO_2 catalyst. *Appl. Catal. A Gen.* **2002**, *236*, 67–76. [CrossRef]
8. Jacobs, G.; Patterson, P.M.; Zhang, Y.Q.; Das, T.; Li, J.L.; Davis, B.H. Fischer–Tropsch synthesis: Deactivation of noble metal-promoted Co/Al_2O_3 catalysts. *Appl. Catal. A Gen.* **2002**, *233*, 215–226. [CrossRef]
9. Martens, J.H.A.; Van't Blik, H.F.J.; Prins, R. Characterization of supported cobalt and cobalt-rhodium catalysts: II. Temperature-Programmed Reduction (TPR) and Oxidation (TPO) of $CoTiO_2$ and $CoRhTiO_2$. *J. Catal.* **1986**, *97*, 200–209. [CrossRef]
10. Lapidus, A.; Krylova, A.; Kazanskii, V.; Borovkov, V.; Zaitsev, A.; Rathousky, J.; Zukal, A.; Jancalkova, M. Hydrocarbon synthesis from carbon monoxide and hydrogen on impregnated cobalt catalysts. Part I. Physico-chemical properties of 10% cobalt/alumina and 10% cobalt/silica. *Appl. Catal.* **1991**, *73*, 65–82. [CrossRef]
11. Diehl, F.; Khodakov, A.Y. Promotion of cobalt Fischer-Tropsch catalysts with noble metals: A review. *Oil Gas Sci. Technol.* **2009**, *64*, 11–24. [CrossRef]
12. Steynberg, A.; Dry, M. *Fischer-Tropsch Technology*; Studies in Surface Science and Catalysis; Elsevier B.V.: Amsterdam, The Netherlands, 2004; Volume 152, ISBN 0-444-51354-X.
13. Den Breejen, J.P.; Radstake, P.B.; Bezemer, G.L.; Bitter, J.H.; Froseth, V.; Holmen, A.; De Jong, K.P. On the origin of the cobalt particle size effects in Fischer-Tropsch catalysis. *J. Am. Chem. Soc.* **2009**, *131*, 7197–7203. [CrossRef] [PubMed]
14. Bezemer, G.L.; Bitter, J.H.; Kuipers, H.P.C.E.; Oosterbeek, H.; Holewijn, J.E.; Xu, X.; Kapteijn, F.; Van Dillen, A.J.; De Jong, K.P. Cobalt Particle Size Effects in the Fischer–Tropsch Reaction Studied with Carbon Nanofiber Supported Catalysts. *J. Am. Chem. Soc.* **2006**, *128*, 3956–3964. [CrossRef] [PubMed]
15. Delannay, F. *Characterization of Heterogeneous Catalysts*; Marcel Dekker Inc.: New York, NY, USA, 1984; ISBN 978-0824771003.
16. Haber, J.; Block, J.H.; Delmon, B. Manual of methods and procedures for catalyst characterization. *Pure Appl. Chem.* **1995**, *67*, 1257–1306. [CrossRef]
17. Rouquerol, F.; Rouquerol, J.; Sing, K. *Adsorption by Powders and Porous Solids—Principles, Methodology and Applications*; Academic Press: Cambridge, MA, USA, 1998; ISBN 978-0-12-598920-6.
18. Arnoldy, P.; Moulijn, J.A. Temperature-programmed reduction of $CoOAl_2O_3$ catalysts. *J. Catal.* **1985**, *93*, 38–54. [CrossRef]
19. Jacobs, G.; Das, T.K.; Zhang, Y.; Li, J.; Racoillet, G.; Davis, B.H. Fischer-Tropsch synthesis: Support, loading, and promoter effects on the reducibility of cobalt catalysts. *Appl. Catal. A Gen.* **2002**, *233*, 263–281. [CrossRef]
20. Borg, O.; Ronning, M.; Storster, S.; Van Beek, W.; Holmen, A. Identification of cobalt species during temperature programmed reduction of Fischer-Tropsch catalysts. *Stud. Surf. Sci. Catal.* **2007**, *163*, 255–272.
21. De la Peña O'Shea, V.A.; Consuelo Álvarez Galván, M.; Platero Prats, A.E.; Campos-Martin, J.M.; Fierro, J.L.G. Direct evidence of the SMSI decoration effect: The case of Co/TiO_2 catalyst. *Chem. Commun.* **2011**, *47*, 7131. [CrossRef] [PubMed]

Article

Biosorption of Methylene Blue Dye Using Natural Biosorbents Made from Weeds

Francisco Silva [1,*], Lorena Nascimento [2], Matheus Brito [3], Kleber da Silva [4], Waldomiro Paschoal Jr. [2,*] and Roberto Fujiyama [1,*]

1 Postgraduate Program in Natural Resource Engineering, Federal University of Pará, Belém, PA 66075-110, Brazil
2 Programa de Pós-Graduação em Física, Universidade Federal do Pará, Belém, PA 66075-110, Brazil
3 Faculty of Chemistry, Federal University of Pará, Belém, PA 66075-110, Brazil
4 Department of Natural Sciences; University of the State of Pará, Belém, PA 66050-540, Brazil
* Correspondence: fxavirlima@gmail.com (F.S.); wpaschoaljr@ufpa.br (W.P.J.); fujiyama.ufpa@gmail.com (R.F.); Tel.: +55-91-98281-1852 (F.S.)

Received: 20 May 2019; Accepted: 11 July 2019; Published: 5 August 2019

Abstract: The purpose of this work is to make use of vegetables that, although widely found in nature, there are few applications. The weeds used here, *Cyanthilium cinereum* (L.) *H. Rob* (CCLHR) and *Paspalum maritimum* (PMT) found in the Amazon region of Belém state of Pará-Brazil, contribute to the problem of water contamination by the removal of the methylene blue dye through the biosorption process, taking advantage of other materials for economic viability and processing. The influences of parameters such as, biosorbent dose, contact time, and initial concentration of dye were examined. The characterizations were realized using SEM to verify the morphology of the material and spectroscopy in the FTIR region. As for the adsorption mechanism, the physical adsorption mechanism prevailed. The time required for the system to reach equilibrium for both biosorbents was from 50 min, following a kinetics described by the pseudo-second order model. The adsorption isotherm data for PMT were better adjusted to the Langmuir model and the biosorption capacity (q_{max}) value was (56.1798 mg/g). CCLHR was better adjusted to the Freundlich model and its maximum biosorption capacity was 76.3359 mg/g. Thus, these weed species are promising for the biosorption of methylene blue dye in effluents.

Keywords: biosorption; weed; methylene blue dye; natural biosorbents; adsorption isotherms; adsorption kinetics

1. Introduction

The occurrence of weeds in the Amazon Region is considered the most serious biological problem faced by cattle ranchers, as well as their control, one of the highest components of the cost of farms production [1]. It is noteworthy that these plants are undesirable and most of the time they are extracted from nature and discarded or eliminated by chemical processes. Another problem in several countries of the world are the industrial processes that generate significant amounts of effluents containing heavy metals and dyes that affect the quality of water one of the resources the most used by living beings. Water is fundamental to the existence and maintenance of life and for this, it must be present in the environment in appropriate quantity and quality [2].

When colorants are present in aquatic environments, color is generally the first impact to be recognized in an effluent because very small amounts of synthetic dyes in water (<1 ppm) are highly visible [3]. This substance causes serious problems of aesthetic nature in receiving water bodies, even when present in small quantities [4]. Dyes, besides affecting the aesthetic value of water bodies, interfere with the penetration of sunlight into the aquatic environment and thus retard photosynthesis, inhibit the growth of aquatic biotics and interfere with the solubility of gases in bodies of water [5]. In the case of effluents from the textile industries, the dye concentration generally ranges from 10 to 200 mg/L, thus being quite visible [6].

The techniques most commonly used in wastewater treatment are reverse osmosis, ion exchange, adsorption, precipitation [7], membrane filtration [8,9], photocatalysis [10–12] and flocculation [13]. Among these methods, adsorption is one of the most effective methods [7,14] and most feasible due to its cost-effective and easy handling [15,16]. Among the adsorbents most applied stand out the zeolites, polymer-based porous materials [17] and, mainly, the activated carbon, most used due to its high surface area, however its use for dye removal is still very expensive, a fact that limits its wide application in the treatment of textile effluents [18]. This has led many researchers to look for more economic and effective adsorbents as potential substitutes for activated carbon [19], resulting in the interest of adsorbents from biomass to be used as sustainable biosorbents.

It is noteworthy that these plants are undesirable and most of the time they are extracted from nature and discarded or eliminated by chemical processes, which is observed two problematic: one is to give a useful end to vegetal species that in the majority of the times causes disorder to different human activities like agricultural, forestry, animal husbandry, ornamental, nautical, energy production between others [20]. The other is chemical contamination of water which is a worldwide concern, in the case here specified by dyes. Compared with other methods, the removal of dyes from aqueous solutions by the adsorption process proved to be an excellent alternative for effluent treatment, as well as an economical technique [21]. The authors report that the use of biological materials for the removal of dyes from aqueous solutions is commonly referred to as biosorption and has now attracted a great deal of interest in scientific knowledge and in the community as sustainable and ecological materials for the production of alternative sorbents. These materials are called biossorvents [22,23].

Aware of the above problems and the search for solutions for the chemical contamination of water, the objective of this research was to evaluate the biosorption potential of biosorbents produced using weeds as *Cyanthilium cinereum* (L.) *H. Rob* (CCLHR) and *Paspalum maritimum Trin* (PMT), collected in the state of Pará amazon region, aiming at the removal of the methylylene blue dye (MB) from aqueous solutions.

The CCLHR and the PMT were characterized by Scanning Electron Microscopy to investigate their morphologies and by Fourier transform infrared spectroscopy (FTIR) to detect the presence of functional groups present in the material that corroborate for their use in the removal of methylene blue (MB) of aqueous solution. PMT is a native species of tropical America, occurring in Central America and the Caribbean, northern Brazil and the coastal zone, from Northeast to South. In Brazil the highest concentrations occur from Pará to Bahia [24]. The CCLHR also known as Vernonia cinerea belongs to the Asteraceae family. The species is native to tropical Africa, tropical Asia, India, Indochina, tropical South America, West India and the US state of Florida [25].

Recent studies indicate that approximately 12% of the synthetic dyes are lost during manufacturing and processing operations and that about 20% end up entering the environment through effluents from industrial wastewater treatment plants [26]. Dyes have a complex chemical structure that is stable to light, heat, oxidizing agents and are also resistant to aerobic digestion [27,28]. Methylene blue is a cationic dye widely used in the textile industry for the dyeing of cotton and wool fabrics. When untreated, uncontrolled discharge into rivers and lakes affect not only the transparency of the waters, but also limits the passage of solar radiation, reducing the natural photosynthetic activity and causing changes in the aquatic biota and causing acute and chronic toxicity of these ecosystems [29,30].

As has already been synthetically mentioned, an appropriate alternative method, which has proved to be quite effective for removal of dyes from aquatic environments is biosorption, a subcategory of adsorption, which uses as biological raw material where the lingnocellulosics are included. In this class of materials, agricultural byproducts have been shown to be efficient because, in addition to being abundant, they are inexpensive and have a relatively low impact on the environment. In comparison to other effluent treatment methods, biosorption substantially reduces the costs associated with financial investment in the process as a whole [31,32].

Several works that deal with the removal of methylene blue are presented in the literature using biosorbents obtained from lignocellulosic materials in the in natura form such as coconut fiber [33], banana peel [34], mint tailings [35], pineapple peel [36], cashew nutshell [37] pine leaves [38], tea residues [39], corn straw, pupunha palm [40], trunk of the papaya tree [41]. In this paper, the biosorbents presented for the removal of methylene blue dye (MB) were produced from weeds that can be defined as any plants that grow spontaneously in a place of human activity and cause damage to this activity, be it agricultural, forestry, livestock, ornamental, nautical, energy production etc. [42].

This work has objective produce biosorbents through weeds (PMT and CCLHR) and realize biosorptions test to verify removal efficiency MB from aqueous solutions. In the biosorption assays, the influence of parameters such as dosage of biosorbents between (0.05–0.5 g), initial concentration of dye in the range of (10.0 and 50.0 mg/L), and contact time (10 and 80 min) were evaluated. The experimental data of biosorption isotherms were evaluated by the Freundlich and Langmuir models and biosorption kinetics by the pseudo-first-order and pseudo-second-order models. The maximum biosorption capacity (q_{max}) values were 56.1798 mg/g for the PMT, whose experimental biosorption isotherm data were better adjusted with the Langmuir model and 76.3359 mg/g for the CCLHR, whose experimental data of biosorption isotherms were better fitted to the Freundlich model. Finally, the research result shows the biosorbent's potential for removal MB from waste water. It is expected that this research will contribute as an alternative in the problematic of the water chemical contamination, using a simple reproduction process and a wide availability of raw material for biosorbents production.

2. Materials and Methods

2.1. Preparation of Biosorbent

The used biosorbent were produced from weeds (*Cyanthilium cinereum* (L.) *H. Rob e Paspalum maritimum Trin*). The weeds found in the Amazon region, Belém-PA, Brazil. They were collected manually within the Universidade Federal do Pará (UFPA). After harvesting, the stems were extracted and cut into lengths of approximately 5 mm. The trimmed stems were washed in distilled water and introduced in an oven at 400 °C for 30 days, to reduce humidity and to avoid attack of microorganisms. After 30 days, the dried samples were again crushed and washed with distilled water until we did not observe more of the coloration in the solution. The wet samples were placed in an oven at 800 °C for 24 h. The prepared biosorbents from weeds were stored in airtight plastic containers in order to avoid humidity, and these were utilized in the biosorption assays.

2.2. Solutions and Reagents

In the present study of biosorption, the used adsorbate was the methylene blue dye (MB), classified as basic or cationic. A stock solution of 100 mg/L was prepared separately, which was diluted to between 10 and 50 mg/L. The molecular structure and chemical formula of the Cationic dye of methylene blue are shown in Table 1.

Table 1. Dye characteristics.

Cationic Dye	Molecular Structure	Chemical Formula
Methylene blue	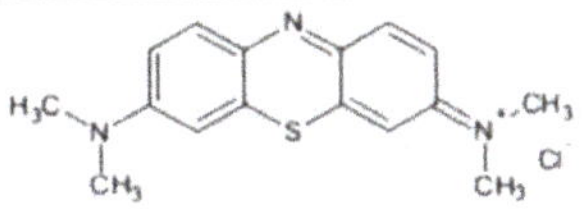 H_3C–N(CH_3)– [phenothiazinium ring: N, S] =N^+(CH_3)CH_3 Cl^-	$C_{16}H_{18}N_3SCl$

2.3. Used Equipments at the Characterization

In the Fourier Transform Infrared Spectroscopy (FT-IR) of the biosorbents was used Perkin Elmer Spectrum Two, in order to verify functional groups present in the samples. The biosorbents morphological was investigated by using Tescan scanning electron microscope (SEM) (Vega3 SB).

2.4. Biosorption Experimental Procedure

The biosorption is biomass ability to adsorb surface pollutants by carboxylic and phenolic functional groups, in which neutral pH make deprotonated and negative charge removes cations from solution by means of process as complexation, ionic exchange and adsorption [43].

The biosorption studies for the assessment of weeds (*Cyanthilium cinereum* (L.) *H. Rob* and *Paspalum maritimum Trin*) for the removal of MB dye from aqueous solutions was conducted by means of the batch biosorption procedure using 50 mL–pH 7 of solution, submitted to constant agitation speed of 150 rpm by magnetic stirrer (QUIMIS–Q221 MAG model), without temperature control. We analyzed the influence of parameters such as biosorbent dosage between 0.05–0.5 g, initial dye concentration in the range of 10.0 and 50.0 mg/L and contact time between 10–80 min. During each procedure at predetermined time intervals, solution samples were taken for residual analysis of MB concentrations using spectrophotometer. Equations (1) and (2) were used to calculate the percentage of removal and the biosorption capacity, respectively:

$$R\% = \frac{(c_0 - c)}{c_0} \times 100\% \quad (1)$$

$$q(t) = \frac{(c_0 - c)}{m} \times \mathrm{V} \quad (2)$$

where:

$R\%$ = Percentage of removal
c_0 = Initial concentration (mg/L)
c = Concentration at time t
V = Volume (L)
$q(t)$ = Biosorption capacity at time t
m = Biosorbent mass.

3. Results

3.1. Characterization of PMT and CCLHR Biosorbents

Figure 1 shows the FTIR results for the identification of the present functional groups in the PMT and CCLHR species. In the range of 3000–3720 cm^{-1}, a broad and low intensity band was observed, peaks at 3330 cm^{-1} (Figure 1a) and 3320 cm^{-1} (Figure 1b), characterizing the presence of O–H (alcohols and phenols) [44]. Peaks were also observed at 1630 cm^{-1} (Figure 1a) and 1613 cm^{-1} (Figure 1b), which indicated the C = O stretching in organic groups of carboxyl and bending vibration of the functional group –OH [45]. The intense band at 1035 cm^{-1} (Figure 1a) and 1033 cm^{-1} (Figure 1b) peaks confirm the functional groups O–C–O of the cellulose and lignin structure.

The biosorption of MB on the adsorbent may be due to the electrostatic attraction between these groups and the cationic dye molecule. At pH above 4, the carboxylic groups are deprotonated and negatively charged carboxylate ligands (–COO–) bind to the positively charged MB molecules. This confirms that the biosorption of MB by adsorbent was an ion exchange mechanism between the negatively charged groups present in adsorbent and the cationic dye molecule [46].

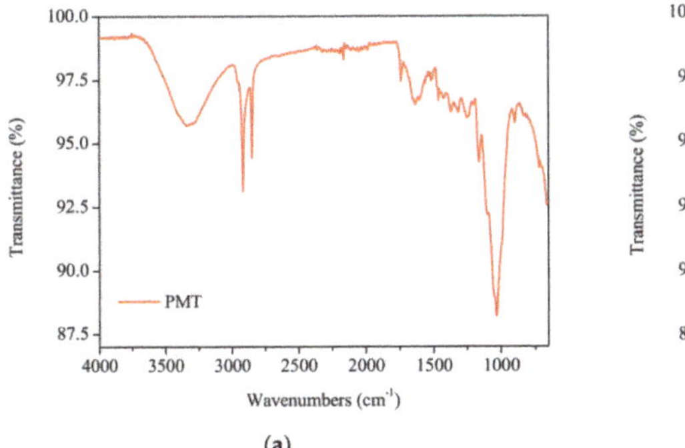

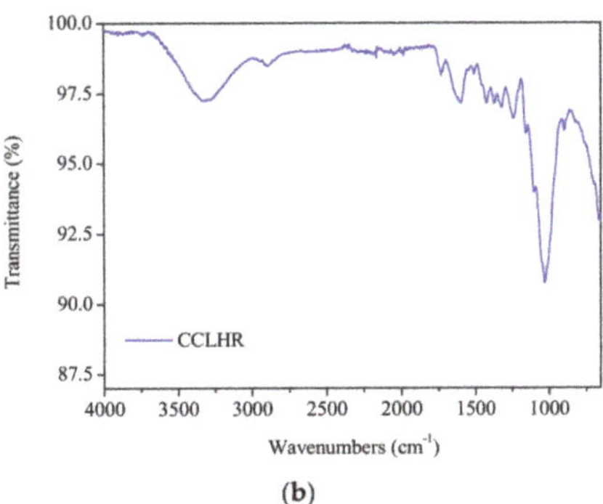

Figure 1. Infrared spectrum of *Paspalum maritimum* (PMT) (**a**) and CCLHR (**b**) samples, in the range of 400–4000 cm^{-1}.

The Scanning Electron Microscopy images of PTM and CCLHR are shown in Figure 2 with magnification of 2190× and 1720×, respectively The SEM analyzes showed a large amount of pores implying a wide surface area, which facilitates the MB biosorption process [40]. This indicates a necessary requirement of these lignocellulosic materials as potential biosorbents. In general, the PMT and CCLHR presented different morphologies along their surface demonstrating different pores sizes and heterogeneous surfaces.

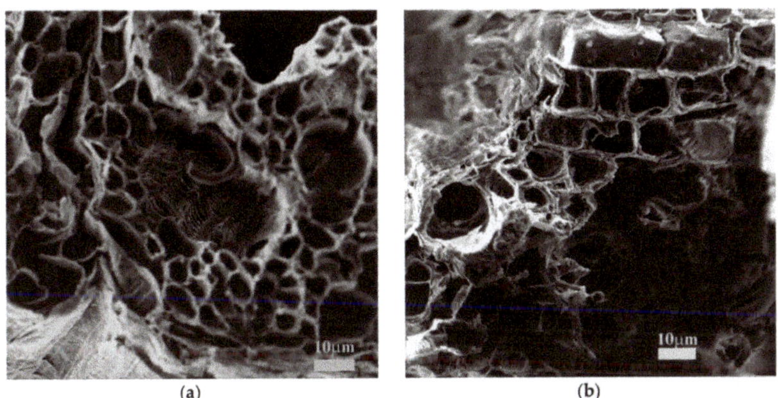

Figure 2. Scanning electron microscope (SEM) image of (**a**) PMT with magnification of 2190× and (**b**) CCLHR with magnification of 1720×.

The characterization results by SEM and FTIR show the predominance the physical adsorption mechanism due to the porosity of the material and the electrostatic interaction between the biosorbent and the methylene blue.

3.2. The Effect of Dosage

The biosorption capacity represent the biosorbate mass can be retained by the biosorbent mass while percentage of removal is related to speed which the biosorbate flows from solution to biosorbent surface. The MB biosorption by the CCLHR and PMT plant species was examined by dosage variation of 0.05 to 0.5 g in the concentration 15 mg/L, solution volume of 50 mL and agitation speed 150 rpm. For both species, we observed that the increase in the mass of produced biosorbents led to an increase at the percentage of removal, with CCLHR from 80.96% to 98.15% and PMT from 92.33% to 95.88%.

However, increasing the dosage from 0.05 to 0.5 g caused a decrease in biosorption capacity, where for CCLHR form 12.11 to 1.47 mg/g, whereas for PMT from 13.85 for 1.44 mg/g. The increase in the percentage of removal with increasing dosage is because the larger amount of mass provided a larger number of active sites available for the biosorption, which causes the increase of the percentage of removal as already reported in the literatures [47,48]. The decrease of the biosorption capacity with the increase of the biosorbent dosage can be explained by the unsaturation of a certain number of active sites, since the volume and concentration remained fixed to a higher mass value, in which should be distributed the same amount of dye. Also, there is the particle aggregation due to the increase of the biosorbent dosage, which causes a decrease in the surface area and the increase of the diffusion path to be travelled by the adsorbate inside the biosorbent [48–54]. Figure 3 shows the curves of biosorbent dose vs. biosorption capacity and biosorbent dose vs. percentage of removal of PTM and CCLHR.

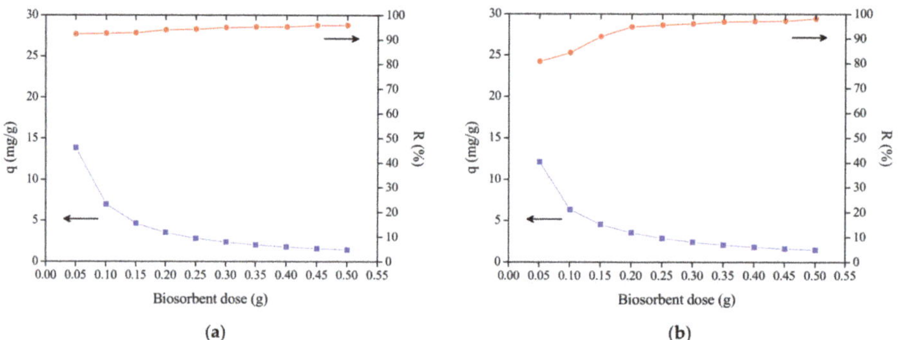

Figure 3. Biosorbent dose vs. biosorption capacity vs. percentage of removal: (**a**) PMT and (**b**) CCLHR.

3.3. *Effect of Initial Dye Concentration*

The initial dye concentration in the range of 10–50 mg/L was studied for the assessment of MB biosorption using a biosorbent dosage of 0.05 g. The biosorption capacity increased with increasing concentration from 9.03 to 41.99 mg/g (MB) for CCLHR biosorbent and from 9.31 to 41.67 mg/g (MB) for PMT biosorbent. The percentage of removal decreased from 90.36% to 83.98% for CCLHR biosorbent and from 93.18% to 83.35% for PTM biosorbent. At lower initial concentrations of MB there are relatively few dye molecules and a large number of available adsorption sites, which are present in the biosorbent masses, thus, it leads to a better interaction of the adsorbate with the biosorption sites, hence resultant to the higher percentage of MB removal. With the increase in the initial concentrations of MB occurs the gradual decrease in the percentage of removal due to the saturation of active biosorbents sites, since with the elevation of concentration the number of the dye molecules increases significantly [55]. The increase of initial dye concentration from 10 mg/L to 50 mg/L provided an increase in the biosorption capacity, since higher concentrations contribute to a decrease in the adsorbate mass transfer resistance of solution to the adsorbent surface, thus filling possible active sites still unoccupied in low concentrations [56,57] (Figure 4).

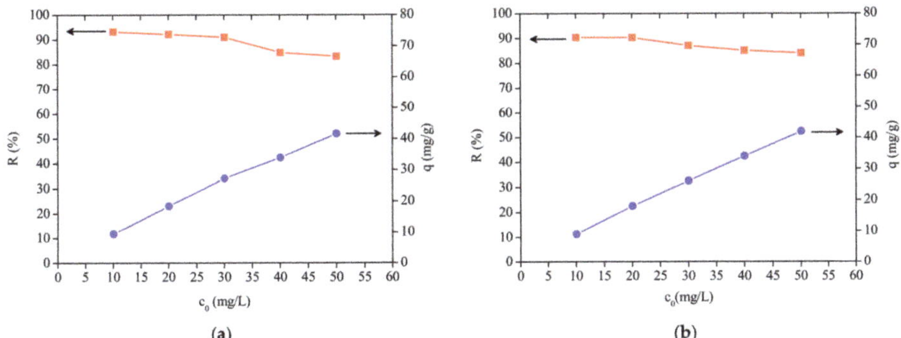

Figure 4. Initial concentration vs. percentage of removal vs. biosorption capacity: (**a**) PMT and (**b**) CCLHR.

3.4. Adsorption Isotherms

Isotherms are diagrams showing the variation of equilibrium concentration of adsorbent with the liquid phase concentration at a temperature. These models are used to illustrate the biosorbent interaction with the biosorbate and provide the relationship between the biosorption capacity and the liquid phase concentration of biosorbate under equilibrium condition at constant temperature [58].

3.4.1. Langmuir Isotherm

The Langmuir model is used in the investigation of dye biosorption from liquid solution [41]. The model based on the assumption that exists a defined number of active sites, the biosorption process occurs on a homogeneous surface through monolayer formation without any interaction with the biosorbed molecules and that all the sites has equivalent energy [56,59,60] The Langmuir isotherm [61] is represented by Equation (3).

$$q_e = \frac{q_{max} k_L c_e}{1 + k_L c_e} \tag{3}$$

Where:

q_e = amount of solute adsorbed per gram of adsorbent at equilibrium (mg/g);
q_{max}: maximum biosorption capacity (mg/g);
k_L interaction constant of adsorbate/adsorbent (L/mg);
c_e: equilibrium concentration of adsorbate (mg/L).

From Equation (3), we can obtain the linearized form:

$$\frac{c_e}{q_e} = \frac{1}{q_{max}} c_e + \frac{1}{k_L q_{max}} \tag{4}$$

And so, plot a chart c_e/q_e in function of c_e which allows you to calculate the which allows you to calculate the q_{max} e k_L being that $1/q_{max}$ is the angular coefficient of the line and the $1/k_L q_{max}$ is the intercession with the ordinate axis.

In Langmuir's model a widely used indicator in terms of analysis is called the separation factor R_L which is calculated on the basis of c_0 e k_L according to Equation (5).

$$R_L = \frac{1}{1 + k_L c_0} \tag{5}$$

c_0 = initial concentration (mg/L).

The value of the constant q_{max} is related to the adsorbed species concentrations on the surface. When the biosorption capacity reaches this value it means that all available sites (sites that the adsorbate binds to the adsorbent) have been filled. The constant k_L is related to the free energy of adsorption, which corresponds to the affinity between the surface of the adsorbent and the adsorbate [58]. R_L indicates the mode of adsorbate biosorbent interaction and allows to classify adsorption isotherms in unfavorable ($R_L > 1$), linear ($R_L = 1$), favorable ($0 < R_L < 1$), or irreversible ($R_L = 0$) [62].

3.4.2. Freundlich Isotherm

According to Freundlich model, the multilayer adsorption occurs on heterogeneous adsorbent surfaces and the higher energy sites on the surface are first occupied and the binding force decreases with the increase in the degree of occupation of the active sites, which reduces the adsorption with time [56,59].

The Freundlich model is represented by [63]:

$$q_e = k_F {c_e}^{\frac{1}{n}} \quad (6)$$

According to Cai L. et al. [44], the linearized form of Equation (6) is described by:

$$\log q_e = \log k_F + \frac{1}{n} \log c_e \quad (7)$$

k_F and n are the Freundlich constants related to biosorption capacity and biosorption intensity, respectively.

The adsorption isotherms obtained experimentally for PMT and CCLHR (Figure 5) were adjusted to the Langmuir and Freundlich models, for initial concentrations in the range between 10 and 50 mg/L. The values of the constants were calculated by the linearized forms of the respective models. Figures 6 and 7 show Freundlich and Langmuir models for both biosorbents (PMT and CCLHR), respectively. The results for the parameters obtained in both biosorbents are shown in Table 2. Analyzing the values of the correlation coefficient in Table 2 and Figure 8, we observed that the experimental data of the adsorption for the PMT is better adjusted to the Langmuir model, while the CCLHR to the Freundlich model. It is justified by the values of the respective linear correlation coefficients (R^2 closer to unity) in each case.

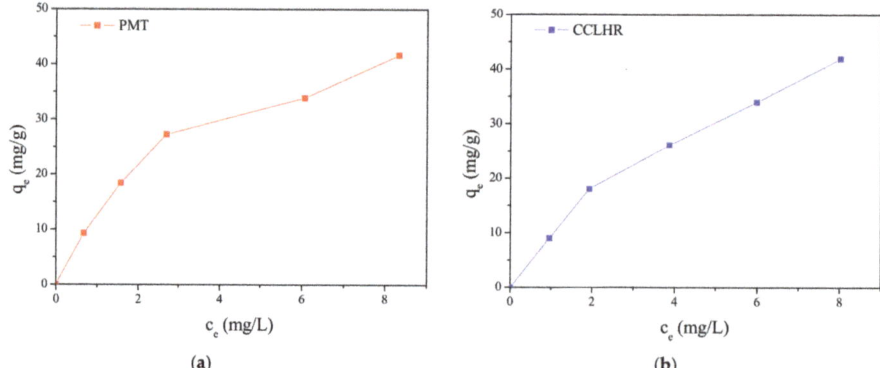

Figure 5. Adsorption isotherm of MB: (**a**) PMT, (**b**) CCLHR.

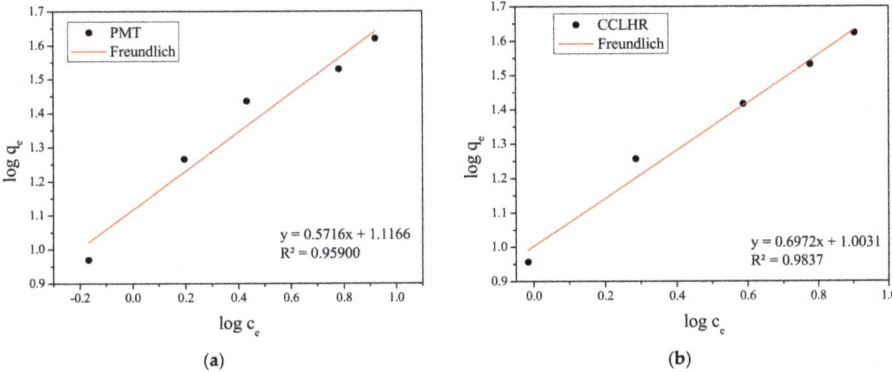

Figure 6. Freundlich isothermal adsorption equation fitting of methylene blue: (**a**) PMT, (**b**) CCLHR.

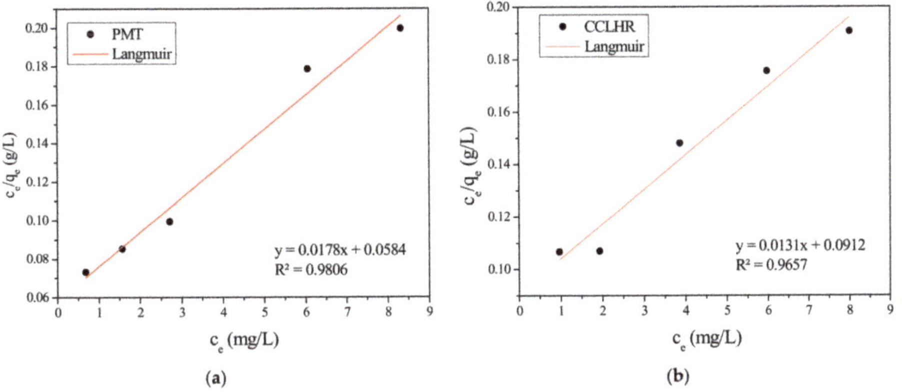

Figure 7. Langmuir isothermal adsorption equation fitting of methylene blue: (**a**) PMT, (**b**) CCLHR.

Table 2. Adsorption isotherm parameters for methylylene blue (MB).

Sample	Freundlich			Langmuir			
	n	k_F	R^2	q_{max} (mg/g)	k_L	R^2	R_L
PMT	1.7495	13.0798	0.9590	56.1798	0.3048	0.9806	0.2470–0.0616
CCLHR	1.4343	10.0716	0.9837	76.3359	0.1436	0.9657	0.4104–0.1222

In relation to the constants k_L, q_{max} and $1/n$ (highest value of n) for the PMT and CCLHR Table 2 we noticed that the highest value of k_L and lower value of $1/n$ were found for the PMT, although the higher value of q_{max} was for CCLHR. Both models indicated higher MB affinity for PMT surface. The higher value of q_{max} for the CCLHR was possibly due to the fact of the greater diffusion in the pores and intraporos of the CCLHR than PMT. Therefore, the CCLHR presented better efficiency in relation to PMT at the conditions investigated in this work.

Various biosorbents have been applied in removal MB from the aqueous solution, as reported in the previous literature, for comparison purposes. We can compare the results with other authors in term the maximum capacity adsorbed in the conditions optimized by each authors: Carica papaya wood (q_{max} = 32.25 mg·g^{-1}) [41], Cornbread (q_{max} = 106.383 mg·g^{-1}) and pupunha palm (q_{max} = 78.989 mg·g^{-1}) [40], Potato shell (q_{max} = 48.7 mg·g^{-1}) [64], Scenedesmus (q_{max} = 61.69 mg·g^{-1}) [65]. Through the comparative study with the Table 2, we can conclude that PMT and CCLHR are between the most efficient adsorbents prepared for industrial wastewater treatment.

The values of R_L between 0 and 1 Langmuir model, whose variation with the initial concentration is shown in Figure 9, for both biosorbents and of n between 1 and 10 Freundlich model are in Table 2, confirm that the biosorption was favorable for both the biosorbents (PMT and CCLHR), ie, the adsorbate prefers the solid phase than liquid [58].

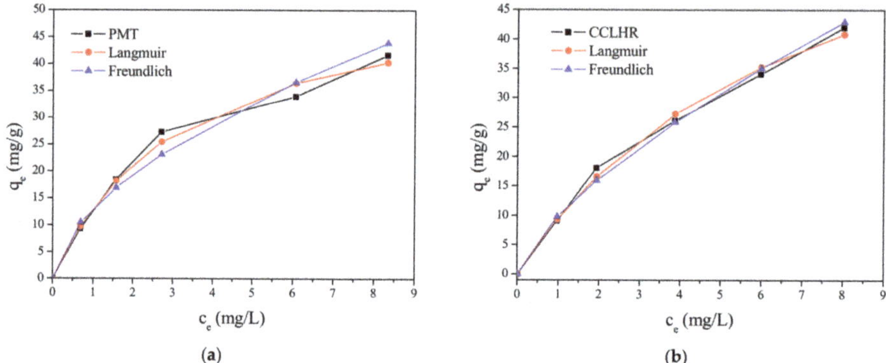

Figure 8. Adsorption isotherms with the Langmuir and Freundlich models: (**a**) PMT, (**b**) CCLHR.

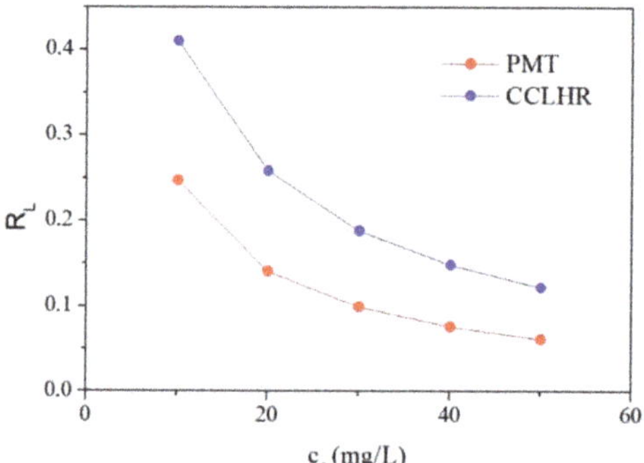

Figure 9. Curve of R_L by the initial concentration for PMT and CCLHR.

3.5. Adsorption Kinetics

The biosorption with the interaction time between the biosorbents and MB was evaluated in the range of 10 to 80 min, a concentration of 15 ppm and a mass of 0.05 g was used. According to Figure 10, we observed that in the first 10 min occurred a rapid increase in the percentage of removal and percentage of biosorption by both biosorbents. After this period, they became slower and remained practically constant from 50 min. The results are due to the fact that in the initial phase of the biosorption, the dyes particles to be biosorbed were almost entirely present in the solution with high probability of accessing the biosorbents surface and the active sites are unoccupied at the beginning of the process. With the increased of the time occurred a decrease in the concentration due to the migration of MB to unoccupied sites, which hindered the biosorption process by increasing the competition of residual dye particles for the remaining available sites. Results in terms of this behavior have already been reported in the literature [59,66–68].

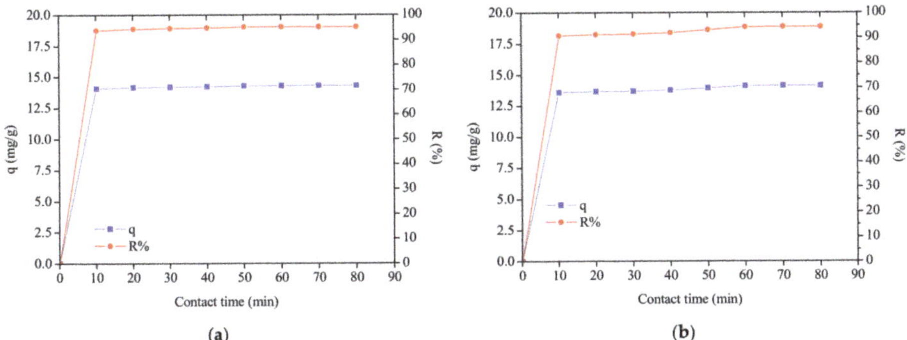

Figure 10. Contact time vs. percentage of removal vs. biosorption capacity: (**a**) PMT and (**b**) CCLHR.

In the literature, there is an expressive amount of linear kinetic models that are used to evaluate the controlling mechanism of the biosorption process such as the chemical reaction, diffusion and mass transfer [69]. Frequently, the most used models are pseudo-first-order and pseudo-second-order, which were also used in this work. The equations and their linearized forms are shown in Table 3.

Table 3. Kinetic equations of biosorption.

Model	Equation	Linearized Formula
Pseudo-first-order	$dq_t/dt = k_1(q_e - q_t)$	$\ln(q_e - q_t) = \ln q_e - k_1 t$
Pseudo-second-order	$dq_t/dt = k_2(q_e - q_t)^2$	$\frac{t}{q_t} = \frac{1}{k_2 q_e^2} + \frac{t}{q_e}$

In order to confirm the experimental data, we utilized the pseudo-first-order and pseudo-second-order models, where kinetic biosorption parameters of MB for initial concentration of 15 ppm are shown in Table 4. In view of these results, it was observed that for both biosorbents the model that best represented the experimental data was the pseudo second order, since R^2 is closer to unity. Figures 11 and 12 show the behavior of the linearized form of said models through which the parameters of Table 4 were calculated.

Table 4. Comparison of the pseudo-first-order and pseudo-second-order models for the biosorption of MB on PMT and CCLHR.

Kinetic Model	Parameter	PMT	CCLHR
Pseudo-first-order	k_1 (mg·min/g)	0.1093	0.1081
	q_e (mg/g)	14.2554	14.1435
	R^2	0.5902	0.5950
Pseudo-second-order	k_2 (mg·min/g)	0.1476	0.0815
	q_e (mg/g)	14.3062	14.2248
	R^2	0.9998	0.9999

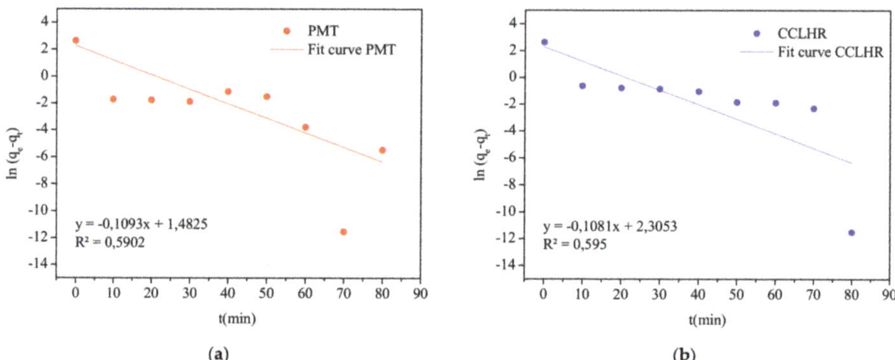

Figure 11. Plots of pseudo-first-order kinetic model for the biosorption: (**a**) PMT and (**b**) CCLHR.

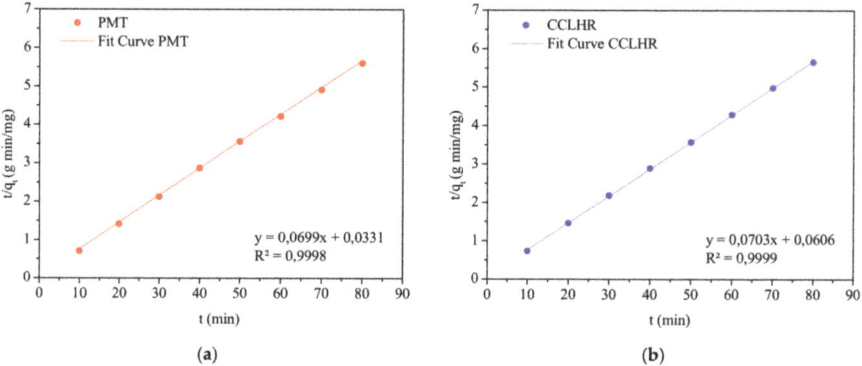

Figure 12. Plots of pseudo-second-order kinetic model for the biosorption: (**a**) PMT and (**b**) CCLHR.

4. Conclusions

The characterizations by FTIR and SEM showed, respectively, functional groups (hydroxyl, carbonyl and carboxyl) and high porosity surface, these factors confirm the produced biosorbents by PMT and CCLHR weeds present appropriate physical-chemical properties to adsorption process. The adsorption kinetics showed high remove percent and experimental data were most be adjusted by pseudo-second-order model. The maximum biosorption capacity (q_{max} = 56.1798 mg·g^{-1} and q_{max} = 76.3359 mg·g^{-1} for PMT and CCLHR, respectively) showed equivalent biosorption value to literature used by MB removal from industries wastewater. The experimental data of the adsorption for the PMT is better adjusted to the Langmuir model, while the CCLHR to the Freundlich model. The range R_L and n results indicated favorable biosorption. Overall, the raw material showed potential for applications with low cost biosorbent, can be a viable alternative and had and with ecological appeal to remove Methylene Blue dyes by several industries segments.

5. Patents

In this work, the used plant raw materials are filed with the National Institute of Industrial Property (INPI)-Brazil, with the code BR 10 2019 008806-0. Additionally, eight weed species divided among the Cyperaceae, Poaceae, Amaranthaceae, Asteraceae families have already been tested and showed biosorption potential.

Author Contributions: Conceptualization, F.S. and R.F.; Methodology, F.S., L.N. and M.B.; Formal analysis, R.F., W.P.J. and K.d.S.; Investigation, F.S., L.N. and M.B.; Resources, F.S. and R.F.; Writing—Original draft, F.S.; Writing—review and editing, R.F., W.P.J. and K.d.S.; Project administration, R.F. and F.S.; Supervision, R.F.

Funding: This research received no external funding.

Acknowledgments: All authors acknowledge financial support from PROPESP/UFPA, CAPES, CNPq. The authors K.S. and W.P.J. gratefully acknowledge support from Universidade Federal do Pará. The author LN gratefully acknowledge support from PIBIC/UFPA for the financial support. In addition, the authors acknowledge the use of the facilities at LABNANO-AMAZON/UFPA.

Conflicts of Interest: The authors declare no conflict of interest.

References

1. Borges, F.C.; Santos, L.S.; Correa, M.J.C. Aleopathic potential of two neolignans isolated from Virola suminanensis (Myristicaceae) leaves. *Planta Daninha* **2007**, *25*, 51–59. [CrossRef]
2. Vidotti, E.C.; Rollemberg, M.C. Algae: From economics in aquatic environments to bioremediation and analytical chemistry. *Quím. Nova* **2004**, *27*, 1–7.
3. Pereira, L.; Alves, M. Dyes—Environmental Impact and Remediation. In *Environmental Protection Strategies for Sustainable Development*; Strategies for Sustainability; Malik, A., Grohmann, E., Eds.; Springer: Dordrecht, The Netherlands, 2012; p. 111.
4. Tundisi, J.G.; Tundisi, T.M. *Limnology*, 1st ed.; Oficina de Textos: São Paulo, Brazil, 2008; p. 632.
5. Souza, N.K. Adsorption of Cationic and Anionic Dyes in Aqueous Solution Using New Bi-Functional Materials of Sugarcane Bagasse. Master's Thesis, Federal University of Ouro Preto, Ouro Preto, Brazil, 2013.
6. Ali, N.; Hameed, A.; Ahmed, S. Physicochemical characterization and bioremediation perspective of textile effluent, dyes and metals by indigenous bacteria. *J. Hazard Mater.* **2009**, *164*, 322–328. [CrossRef] [PubMed]
7. Gupta, V.K.; Suhas Ali, I.; Saini, V.K. Removal of rhodamine B, fast green, and methylene blue from wastewater using red mud, an aluminum industry waste. *Ind. Eng. Chem. Res.* **2004**, *43*, 1740–1747. [CrossRef]
8. Yamashita, M.S.J.; Kikutani, T.; Hashimoto, T. Activated carbon fibers and films derived from poly (vinylidene fluoride). *Carbon* **2001**, *39*, 207–214. [CrossRef]
9. Hu, D.D.; Lin, J.; Zhang, Q.; Lu, J.N.; Wang, X.Y.; Wang, Y.W.; Bu, F.; Ding, L.F.; Wang, L.; Wu, T. Multi-step host–guest energy transfer between inorganic chalcogenide-based semiconductor zeolite material and organic dye molecules. *Chem. Mater.* **2015**, *27*, 4099–4104. [CrossRef]
10. Shang, L.; Bian, T.; Zhang, B.; Zhang, D.; Wu, L.Z.; Tung, C.H.; Yin, Y.; Zhang, T. Graphene-supported ultrafine metal nanoparticles encapsulated by mesoporous silica: robust catalysts for oxidation and reduction reactions, Angew. *Chem. Int. Ed.* **2014**, *53*, 250–254. [CrossRef] [PubMed]
11. Hossaini, H.; Moussavi, G.; Farrokhi, M. The investigation of the LED-activated FeFNS-TiO_2 nanocatalyst for photocatalytic degradation and mineralization of organophosphate pesticides in water. *Water Res.* **2014**, *59*, 130–144. [CrossRef] [PubMed]
12. Echavia, G.R.; Matzusawa, F.; Negishi, N. Photocatalytic degradation of organophosphate and phosphonoglycine pesticides using TiO_2 immobilized on silica gel. *Chemosphere* **2009**, *76*, 595–600. [CrossRef]
13. Ahmed, S.M.; El-Dib, F.I.; El-Gendy, N.; Sayed, W.M.; El-Khodary, M. A kinetic study for the removal of anionic sulphonated dye from aqueous solution using nano-polyaniline and Baker's yeast. *Arab. J. Chem.* **2016**, *9*, S1721–S1728. [CrossRef]
14. Dabrowski, A. Adsorption e from theory to practice. *Adv. Colloid Interface Sci.* **2001**, *93*, 135–224. [CrossRef]
15. Fan, L.; Luo, C.; Sun, M.; Qiu, H.; Li, X. Synthesis of magnetic beta-cyclodextrin-chitosan/graphene oxide as nanoadsorbent and its application in dye adsorption and removal. *Colloids Surf. B* **2013**, *103*, 601–607. [CrossRef] [PubMed]
16. Guo, H.; Lin, F.; Chen, J.; Li, F.; Weng, W. Metal-organic framework MIL-125(Ti) for efficient adsorptive removal of Rhodamine B from aqueous solution. *Appl. Organomet. Chem.* **2015**, *29*, 12–19. [CrossRef]
17. Cui, Y.-Y.; Zhang, J.; Ren, L.-L.; Cheng, A.-L.; Gao, E.-Q. A functional anionic metal–organic framework for selective adsorption and separation of organic dyes. *Polyhedron* **2018**, *161*, 71–77.
18. Gupta, V.K.; Suhas. Application of low-cost adsorbents for dye removal—A review. *J. Environ. Manag.* **2009**, *90*, 2313–2342. [CrossRef] [PubMed]

19. Gusmão, G.K.A.; Gurgel, A.L.V.; Melo, S.T.M.; Gil, L.F. Application of succinylated sugarcane bagasse as adsorbent to remove methylene blue and gentian violet from aqueous solutions—Kinetic and equilibrium studies. *Dyes Pigment.* **2012**, *92*, 967–974. [CrossRef]
20. Calvete, T. Pinatura in Natura and Active Coal Shells—Adsorbents for Dye Removal in Aqueous Effluents. Ph.D. Thesis, Chemistry, Federal University of Rio Grande do Sul, Porto Alegre, Brazil, 2011.
21. Crini, G. Non-conventional low-cost adsorbents for dye removal: A review. *Bioresour. Technol.* **2006**, *97*, 1061–1085. [CrossRef]
22. Rafatullah, M.; Sulaiman, O.; Hashim, R.; Ahmad, A. Adsorption of methylene blue on low-cost adsorbents: A review. *J. Hazard. Mater.* **2010**, *177*, 70–80. [CrossRef] [PubMed]
23. Boudechiche, N.; Mokaddem, H.; Sadaoui, Z.; Trari, M. Biosorption of cationic dye from aqueous solutions onto lignocellulosic biomass (*Luffa cylindrica*): Characterization, equilibrium, kinetic and thermodynamic studies. *Int. J. Ind. Chem.* **2016**, *7*, 167–180. [CrossRef]
24. Kissmann, K.G. *Weed and Noxious Plants*, 1st ed.; BASF: São Paulo, Brazil, 1991; pp. 1–126.
25. Swetha Bindu, C.H.; Prathibha, B. Evaluation of antioxidant activity of ethanolic extrat of leaves of *Cyanthilium cinereum* (L). H. Rob. by using isolated frog heart. *IJPPR Hum.* **2018**, *12*, 458–465.
26. Essawy, A.A.; Ali, A.E.-H.; Abdel-Mottaleb, M.S.A. Application of novel copolymer-TiO_2 membranes for some textile dyes adsorptive removal from aqueous solution and photocatalytic decolorization. *J. Hazard. Mater.* **2008**, *157*, 547–552. [CrossRef]
27. Sun, Q.Y.; Yang, L.Z. The adsorption of basic dyes from aqueous solution on modified peat-resin particle. *Water Res.* **2003**, *37*, 1535–1544. [CrossRef]
28. Gupta, V.K.; Mittal, A.; Krishnan, L.; Gajbe, V. Adsorption kinetics and column operations for the removal and recovery of malachite green from wastewater using bottom ash. *Sep. Purif. Technol.* **2004**, *40*, 87–96. [CrossRef]
29. Kunz, A.; Zamora, P.P.; Moraes, S.G.; Durán, N. New trends in effluent treatment of textile effluents. *Quím. Nova* **2002**, *25*, 78–82. [CrossRef]
30. Silva, W.L.L.D.; Oliveira, S.P.D. Modification of the adsorption characteristics of the sugarcane bagasse for the removal of methylene blue from aqueous solutions. *Sci. Plena* **2012**, *8*, 1–9.
31. Abdolali, A.; Guo, W.S.; Ngo, H.H.; Chen, S.S.; Nguyen, N.C.; Tung, K.L. Typical lignocellulosic wastes and by-products for biosorption process in water and wastewatertreatment: A critical review. *Bioresour. Technol.* **2014**, *160*, 57–66. [CrossRef]
32. Vijayaraghavan, K.; Rangabhashiyam, S.; Ashokkumar, T.; Arockiaraj, J. Assessment of samarium biosorption from aqueous solution by brown macroalga *Turbinaria conoides*. *J. Taiwan Inst. Chem. E* **2017**, *74*, 113–120. [CrossRef]
33. Wong, Y.C.; Senan, M.S.R.; Atiqah, N.A. Removal of methylene blue and malachite green dye using different form of coconut fibre as absorbent. *J. Basic Appl. Sci.* **2013**, *9*, 172–177. [CrossRef]
34. Mourarak, F.; Atmani, R.; Maghri, I.; Elkouali, M.; Talbi, M.; Bouamrani, M.L.; Salouhi, M.; Kenz, A. Elimination of methylene blue dye with natural adsorbent-banana peels powder. *GJSFR* **2014**, *14*, 38–44.
35. Ainane, T.; Khammour, F.; Belghazi, O.; Kabbaj, M.; Yousfi, S.; Talbi, M.M. Elkouali. Study and modelling of kinetics biosorption of methylene blue on biomass material from waste mint. *Biotechnology* **2015**, *11*, 281–285.
36. Weng, C.-H.; Lin, Y.-T.; Tzeng, T.-W. Removal of methylene blue from aqueous solution by adsorption onto pineapple leaf powder. *J. Hazard. Mater.* **2009**, *170*, 417–424. [CrossRef]
37. Senthil Kumar, P.; Abhinaya, R.V.; Gayathri Lashmi, K.; Arthi, V.; Pavithra, R.; Sathyaselvabala, V.; Sivanesan, S. Adsorption of methylene blue dye from aqueous solution by agricultural waste: Equilibrium, thermodynamics, kinetics, mechanism and process design. *Colloid. J.* **2011**, *73*, 651–661. [CrossRef]
38. Yagub, M.T.; Sen, T.K.; Ang, H.M. Equilibrium, kinetics, and thermodynamics of methylene blue adsorption by Pine tree leaves. *Water Air Soil Pollut.* **2012**, *223*, 5267–5282. [CrossRef]
39. Uddin, M.T.; Islam, M.A.; Mahmud, S.; Rukanuzzaman, M. Adsorptive removal of methylene blue by tea waste. *J. Hazard. Mater.* **2009**, *164*, 53–60. [CrossRef]
40. Honorato, A.C.; Machado, J.M.; Celante, G.; Borges, W.G.P.; Dragunski, D.C.; Caetano, J. Biosorption of methylene blue using agroindustrial residues. *Rev. Bras. Eng. Agric. Amb.* **2015**, *19*, 705–710. [CrossRef]
41. Rangabhashiyam, S.; Lata, S.; Balasubramanian, P. Biosorption characteristics of methylene blue and malachite green from simulated wastewater onto *Carica papaya* wood biosorbent. *Surf. Interfaces* **2018**, *10*, 197–215.
42. Carvalho, L.B. *Weeds*, 1st ed.; Leonardo Bianco de Carvalho: Lages, Brazil, 2013; p. 82.

43. Pimentel, A.M.R. Removal of Co (II) and Mn (II) from Aqueous Solutions Using the Biomass *R. opacus*. Master's Thesis, PUC—Rio, Rio de Janeiro, Brazil, 2011.
44. Cai, L.; Zhang, Y.; Zhou, Y.; Zhang, X.; Ji, L.; Song, W.; Liu, J. Effective adsorption of diesel oil by Crab-Shell-Derived biochar nanomaterials. *Materials* **2019**, *12*, 236. [CrossRef]
45. Zhang, X.; Hao, Y.; Wang, X.; Chen, Z. Rapid removal of zinc(II) from aqueous solutions using a mesoporous activated carbon prepared from agricultural waste. *Materials* **2017**, *10*, 1002. [CrossRef]
46. Saeed, A.; Sharif, M.; Iqbal, M. Application potential of grapefruit peel as dye sorbent: Kinetics, equilibrium and mechanism of crystal violet adsorption. *J. Hazard. Mater.* **2010**, *179*, 564–572. [CrossRef]
47. Alencar, W.S.; Acayanka, E.; Lima, E.C.; Royer, B.; de Souza, F.E.; Lameira, J.; Alves, C.N. Application of *Mangifera indica* (mango) seeds as a biosorbent for removal of Victazol Orange 3R dye from aqueous solution and study of the biosorption mechanism. *Chem. Eng. J.* **2012**, *209*, 577–588. [CrossRef]
48. Tural, B.; Ertaş, E.; Enez, B.; Fincan, S.A.; Tural, S. Preparation and characterization of a novel magnetic biosorbent functionalized with biomass of Bacillus Subtilis: Kinetic and isotherm studies of biosorption processes in the removal of Methylene Blue. *J. Environ. Chem. Eng.* **2017**, *5*, 4795–4802. [CrossRef]
49. Royer, B.; Cardoso, N.F.; Lima, E.C.; Ruiz, V.S.O.; Macedo, T.R.; Airoldi, C. Organofunctionalized kenyaite for dye removal from aqueous solution. *J. Colloid Interface Sci.* **2009**, *336*, 398–405. [CrossRef]
50. Royer, B.; Cardoso, N.F.; Lima, E.C.; Macedo, T.R.; Airoldi, C. A useful organofunctionalized layered silicate for textile dye removal. *J. Hazard. Mater.* **2010**, *181*, 366–374. [CrossRef]
51. Da Silva, L.G.; Ruggiero, R.; Gontijo, P.M.; Pinto, R.B.; Royer, B.; Lima, E.C.; Calvete, T. Adsorption of Brilliant Red 2BE dye from water solutions by a chemically modified sugarcane bagasse lignin. *Chem. Eng. J.* **2011**, *168*, 620–628. [CrossRef]
52. Cardoso, N.F.; Pinto, R.B.; Lima, E.C.; Calvete, T.; Amavisca, C.V.; Royer, B.; Pinto, I.S. Removal of remazol black B textile dye from aqueous solution by adsorption. *Desalination* **2011**, *269*, 92–103. [CrossRef]
53. Calvete, T.; Lima, E.C.; Cardoso, N.F.; Dias, S.L.P.; Pavan, F.A. Application of carbon adsorbents prepared from the Brazilian pine-fruit-shell for the removal of Procion Red MX 3B from aqueous solution—Kinetic, equilibrium, and thermodynamic studies. *Chem. Eng. J.* **2009**, *155*, 627–636. [CrossRef]
54. Calvete, T.; Lima, E.C.; Cardoso, N.F.; Vaghetti, J.C.P.; Dias, S.L.P.; Pavan, F.A. Application of carbon adsorbents prepared from Brazilian-pine fruit shell for the removal of reactive orange 16 from aqueous solution: Kinetic, equilibrium, and thermodynamic studies. *J. Environ. Manag.* **2010**, *91*, 1695–1706. [CrossRef]
55. Hassan, W.; Farooq, U.; Ahmad, M.; Athar, M.; Khan, M.A. Potential biosorbent, Haloxylon recurvum plant stems, for the removal of methylene blue dye. *Arab. J. Chem.* **2017**, *10*, 1512–1522. [CrossRef]
56. Miraboutalebi, S.M.; Nikouzad, S.K.; Peydayesh, M.; Allahgholi, N.; Vafajoo, L.; McKay, G. Methylene blue adsorption via maize silk powder: Kinetic, equilibrium, thermodynamic studies and residual error analysis. *Process Saf. Environ. Prot.* **2017**, *106*, 191–202. [CrossRef]
57. Leal, P.V.B.; Gregório, A.M.; Otoni, F.; da Silva, P.R.; de Krauser, M.O.; Holzbach, J.C. Study of adsorption of methylene blue dye on babassu. *J. Biotec. Biodiv.* **2012**, *3*, 166–171.
58. Nascimento, R.F.; Lima, A.C.A.; Vidal, C.B.; Melo, D.Q.; Raulino, G.S.C. *Adsorption: Theoretical Aspects and Environmental Applications*, 1st ed.; UFC: Fortaleza, Brazil, 2014; p. 256.
59. Tang, Y.; Zeng, Y.; Hu, T.; Zhou, Q.; Peng, Y. Preparation of lignin sulfonate-based mesoporous materials for adsorbing malachite green from aqueous solution. *J. Environ. Chem. Eng.* **2016**, *4*, 2900–2910. [CrossRef]
60. Langmuir, I. The adsorption of gases on plane surfaces of glass, mica and platinum. *J. Am. Chem. Soc.* **1918**, *40*, 1361–1403. [CrossRef]
61. Limousin, G.; Gaudet, J.-P.; Charlet, L.; Szenknect, S.; Barthès, V.; Krimissa, M. Sorption isotherms: A review on physical bases, modeling and measurement. *Appl. Geochem.* **2007**, *22*, 249–275. [CrossRef]
62. Pal, S.; Ghorai, S.; Das, C.; Samrat, S.; Ghosh, A.; Panda, A.B. Carboxymethyl tamarind-g-poly (acrylamide)/silica: A high performance hybrid nanocomposite for adsorption of methylene blue dye. *Ind. Eng. Chem. Res.* **2012**, *51*, 15546–15556. [CrossRef]
63. Freundlich, H.M. Over the adsorption in solution. *J. Phys. Chem.* **1906**, *57*, 385–470.
64. Alfredo, A.P.C.; Gonçalves, G.C.; Lobo, V.S.; Montanher, S.F. Adsorção de azul de metileno em casca de batata utilizando sistemas em batelada e coluna de leito fixo. *Rev. Virtual Quim.* **2015**, *7*, 1909–1920. [CrossRef]

65. Afshariani, F.; Roosta, A. Experimental study and mathematical modeling of biosorption of methylene blue from aqueous solution in a packed bed of microalgae Scenedesmus. *J. Clean. Prod.* **2019**, *225*, 133–142. [CrossRef]
66. Kumar, A.; Jena, H.M. Removal of methylene blue and phenol onto prepared activated carbon from Fox nutshell by chemical activation in batch and fixed-bed column. *J. Clean. Prod.* **2016**, *137*, 1246–1259. [CrossRef]
67. Pang, J.; Fu, F.; Ding, Z.; Lu, J.; Li, N.; Tang, B. Adsorption behaviors of methylene blue from aqueous solution on mesoporous birnessite. *J. Taiwan Inst. Chem. E* **2017**, *77*, 168–176. [CrossRef]
68. Heidarinejad, Z.; Rahmanian, O.; Fazlzadeh, M.; Heidari, M. Enhancement of methylene blue adsorption onto activated carbon prepared from Date Press Cake by low frequency ultrasound. *J. Mol. Liq.* **2018**, *264*, 591–599. [CrossRef]
69. Da Rocha, O.R.S.; do Nascimento, G.E.; Campos, N.F.; da Silva, V.L.; Duarte, M.M.M.B. Evaluation of the adsorptive process using green coconut mesocarp for the removal of the reactive gray dye BF-2R. *Quím. Nova* **2012**, *35*, 1369–1374.

MDPI
St. Alban-Anlage 66
4052 Basel
Switzerland
Tel. +41 61 683 77 34
Fax +41 61 302 89 18
www.mdpi.com

Materials Editorial Office
E-mail: materials@mdpi.com
www.mdpi.com/journal/materials

www.ingramcontent.com/pod-product-compliance
Lightning Source LLC
LaVergne TN
LVHW070635170726
843515LV00004B/251
* 9 7 8 3 0 3 9 3 6 2 7 4 5 *